NONLINEAR FEEDBACK CONTROL SYSTEMS

An Operator Theory Approach

NONLINEAR FEEDBACK CONTROL SYSTEMS

An Operator Theory Approach

Rui J. P. de Figueiredo
Department of Electrical & Computer Engineering
and Department of Mathematics
University of California at Irvine
Irvine, California

Guanrong Chen
Department of Electrical Engineering
University of Houston
Houston, Texas

ACADEMIC PRESS, INC.
Harcourt Brace & Company, Publishers

Boston   San Diego   New York
London   Sydney   Tokyo   Toronto

Copyright © 1993 by Academic Press, Inc.
All rights reserved.
No part of this publication may be reproduced or
transmitted in any form or by any means, electronic
or mechanical, including photocopy, recording, or
any information storage and retrieval system, without
permission in writing from the publisher.

ACADEMIC PRESS, INC.
1250 Sixth Avenue, San Diego, CA 92101-4311

United Kingdom edition published by
ACADEMIC PRESS LIMITED
24–28 Oval Road, London NW1 7DX

Library of Congress Cataloging-in-Publication Data
DeFigueiredo, Rui J. P.
    Nonlinear feedback control systems: an operator theory approach /
  Rui J. P. de Figueiredo and Guanrong Chen
        p. cm.
    Includes bibliographical references and index.
    ISBN 0-12-208630-9
    1. Feedback control systems. 2. Linear systems. I. Chen, G.
  (Guanrong) II. Title.
  TJ216.D384   1993
  629.8'3—dc20   93-19434
                                                                        CIP

Printed in the United States of America
93 94 95 96   BB   9 8 7 6 5 4 3 2 1

Contents

Preface

The progress made in multivariable control theory, especially over the last decade, has enabled the *linear* part of this theory to achieve a high degree of maturity. For *nonlinear* systems, however, the need still remains for theoretical developments that may continue to motivate and tie up the emerging results. This need is accentuated by the enormous computational power that is becoming available to control theorists and engineers. Computer-oriented approaches to nonlinear control systems analysis and design are placing an increasing demand for a mathematical theory that will facilitate the formulation of the inherently difficult nonlinear problems, the understanding of the underlying system properties, and the construction of the pertinent algorithms.

Motivated by these considerations, we have written the present book in order to introduce control systems analysts and engineers to the analysis of nonlinear feedback control systems based on an operator-theoretic approach.

We may point out in this connection that, in addition to the "classical" approach, which is based on conventional real and complex variable theory, there have been mainly two rapidly developing general approaches to the field: the differential-geometric approach and the operator-theoretic approach. On the differential-geometric approach, several good textbooks and monographs have appeared recently. On the other hand, there seems to be a need for an introductory treatise for the operator-theoretic approach for use by newcomers in the field. We hope that the present book will satisfy this purpose. In order to make the presentation self-contained and self-unified, and yet keep the length of the book within acceptable bounds, we have restricted the material to fit our own research interests and points of view.

The book emphasizes analysis rather than design because we feel that developments in nonlinear analysis have reached maturity while those in nonlinear design are still in the growth phase. The analysis that we have presented addresses qualitative properties and as such should provide significant insights to assist in systems

vii

design and synthesis. The design principles that we have presented in detail address problems such as optimal tracking, robust stabilization, and disturbance rejection.

The first two chapters provide the mathematical background for the remaining four chapters. They are basic enough to serve as a foundation for other research topics in systems theory. More precisely, we present the so-called n-linear operator theory in a unified way in Chapter 1, all the related materials on which can only be found in scattered research papers, survey articles, or perhaps books not published in English. This chapter closes with a treatment of general nonlinear analytic operator theory, including the nonlinear Volterra operators, which have been widely used in the area of nonlinear systems analysis and control engineering. Chapter 2 is devoted to the classical nonlinear Lipschitz operator theory and the generalized nonlinear Lipschitz operator theory that was developed by the present authors in the last few years. These theories are interesting in their own right, not to mention their important applications to nonlinear systems control treated in the last few chapters of the book.

Based on the operator theory developed as above, Chapter 3 establishes a general framework for the analysis of nonlinear feedback control systems. In particular, the main issue of stability and stabilization of a nonlinear feedback control system is addressed and discussed in detail.

Chapter 4 describes some basic design problems, such as system output tracking, robust stabilization, and disturbance rejection of nonlinear control systems under certain optimality criteria. Motivation for these problems is provided by an example from robotics. The chapter focuses on nonlinear feedback design strategies and principles rather than on technical implementation issues. Chapter 5 is devoted to the special topic of the coprime factorization problem for nonlinear feedback control systems. Coprime factorizations, especially right coprime factorizations, of nonlinear mappings that describe feedback control systems have been a very interesting topic that has attracted much attention recently due to their usefulness in stabilization of nonlinear systems. Almost all the materials discussed in this chapter have been developed by the present authors, including a necessary and sufficient condition for the existence of a right coprime factorization and construction methods based on the generalized Lipschitz operator theory. The last chapter of the book discusses the nonlinear system identification problem, in the form of two issues. The first issue addressed is the complexity problem in nonlinear system identification, and the second is the identification of a nonlinear plant in the form of a general Nth-order nonlinear difference equation, using a generalized Fock space framework. These results and techniques, developed mainly by the present authors, are expected to have many applications in related areas in the near future.

The material in this book is relatively self-contained and written in a format that is consistent throughout the text. We assume that the reader has some background in real and functional analyses such as elementary Banach spaces and linear operator theory. No differential geometry, abstract algebra, or spectral analysis is required for the understanding of the material presented. In each chapter, references are given only to those publications that are directly related to the contents and to those from which some materials have been taken (modified) and used in the text. In order to simplify the presentation we did not refer to many valuable sources, as usually done in other books. In case we missed some important references which we should not have missed, we apologize in advance to the authors of those references.

We have a remark on the terminology. We frequently mix the use of a "mapping" and an "operator," only for linguistic reasons. Since we did not consider any set-valued functions in this book, the alternative use of the two names with the same meaning should not cause any confusion in one's reading.

Finally, we would like to thank Professors Charles Desoer, Vaclav Dolezal, Jacob Hammer, Chyi Hwang, Greg Wasilkowski, and Yaoqi Yu for helpful discussions on the topics in this book.

Rui J. P. de Figueiredo
Guanrong Chen

Chapter 1

Nonlinear Analytic Mappings

This chapter will be devoted to developing a basic theory of nonlinear analytic mappings defined from a normed linear space to a Banach space of complex-valued functions. Although it is possible to extend the results of this chapter to more abstract settings, we only restrict our attention to complex-valued functions defined on the time domain, for the purpose of engineering applications to be further explored later. Nonlinear analytic mappings constitute a natural generalization of the concept of classical analytic functions in complex variables to certain bounded nonlinear operators. The notion of nonlinear analytic mappings to be studied in this chapter includes the n-linear and n-power operators defined from one linear space to another; the boundedness, continuity, and differentiability of the n-linear and n-power operators defined on normed linear spaces; the absolutely and uniformly convergent infinite power series expansions of some bounded nonlinear operators mapping from a normed linear space to a Banach space; and the mathematical description of nonlinear analytic mappings, which contain the well-known and elegant nonlinear Volterra series representations of certain bounded nonlinear operators used for years in systems control and signal processing.

1

§ 1. n-Linear Operators

Let X and Y be linear spaces of complex-valued functions on the real line such that for any $x_1, \ldots, x_n \in X$, the tensor product of $x_1, x_2, \ldots, x_n$, denoted by $x_1 x_2 \cdots x_n$, is a well-defined element in Y. For example, in the case that $X = L_2$, the standard Hilbert space of square-integrable functions, since the usual product $x_1 \cdots x_n$ of the functions $x_1, \cdots, x_n \in L_2$ is well defined and belongs to L_2, Y can be chosen to be the same space: $Y = X = L_2$. Note, however, that a tensor product can be much more general. In what follows, we will denote by X^n the n-product space of X, namely:

$$X^n = X \times \cdots \times X.$$

An element of X^n will be denoted by $\mathbf{x} = (x_1, \ldots, x_n) \in X^n$, where $x_k \in X$ for all $k = 1, \ldots, n$. X^n is a linear space if the linear operations between any two elements of X^n are performed coordinatewise, in the sense that all the other coordinates are fixed.

1.1. Definitions. An operator $f_n : X^n \to Y$ is said to be *n-linear* if f_n is linear with respect to each component $x_k : k = 1, \ldots, n$, in the sense that f_n is a linear operator with respect to x_k for each k while all the other variables x_ℓ with $\ell \neq k$ are fixed. In notation, we write

$$f_n : (x_1, \ldots, x_n) \to f_n(x_1, \ldots, x_n)$$

for an n-linear operator $f_n(\cdot, \ldots, \cdot)$. If, furthermore,

$$f_n(x_1, \ldots, x_n) = f_n(x_{\ell_1}, \ldots, x_{\ell_n})$$

for any permutation $\{x_{\ell_1}, \ldots, x_{\ell_n}\}$ of $\{x_1, \ldots, x_n\}$, then f_n is called a *symmetric n-linear operator*.

In particular, for $n = 2$, $f_2 : (x_1, x_2) \to f_2(x_1, x_2)$ is called a bilinear operator and, using such bilinear operators,

$$\tilde{f}_2 : (x_1, x_2) \to f_2(x_1, x_2) + f_2(x_2, x_1)$$

defines a symmetric bilinear operator.

1.2. Examples. The following two operators f_n and g_n are n-linear:

$$f_n : (x_1, \ldots, x_n) \to c x_1 \cdots x_n \,,$$

$$g_n : (x_1, \ldots, x_n) \to c x_1 x_n \,,$$

where $c \in \mathcal{C}$ and $x_1, \ldots, x_n \in L_2$. However, f_n is symmetric, while g_n is not if $n \geq 3$.

Some other important and useful n-linear operators will be shown below in Section 9 of this chapter.

It is easily seen that an n-linear operator is nonlinear in the usual sense if $n \geq 2$, except for some very special cases such as the projection operator p_n defined by

$$p_n : (x_1, \ldots, x_n) \to x_n \,.$$

1.3. Definitions. Let f_n be an n-linear operator mapping from X^n to Y. If, furthermore, f_n satisfies that

$$f_n : (x, \ldots, x) \to c_n x^n \,, \quad \text{for some } c_n \in \mathcal{C} \,,$$

in the usual sense of product for all functions $x \in X$, then f_n is said to be a *homogeneous polynomic operator of degree n*, or simply an *n-power operator*. More generally, if

$$f_{k,n-k} : (x, \ldots, x, h, \ldots, h) \to c_n x^k h^{n-k} \,, \quad \text{for some } c_n \in \mathcal{C} \,,$$

for all $\mathbf{x} = (x, \ldots, x, h, \ldots, h) \in X^n$ with k components in x and $(n-k)$ components in h, then f_n is said to be a $(k, n-k)$-*power operator*.

In notation, we sometimes simply write $f_n : X \to Y$, or more precisely,

$$f_n : (x, \ldots, x) \to f_n[x^n]$$

for an n-power operator $f_n[\cdot]$ when its domain is restricted to the diagonal of the product space X^n. The reason is that each coordinate axis of the diagonal of the product space X^n is identical to X, so that an n-power operator $f_n[\cdot]$ can also be considered to be defined on X. Similarly, we write $f_{k,n-k} : X \times X \to Y$ or

$$f_{k,n-k} : (x, \ldots, x, h, \ldots, h) \to f_{k,n-k}[x^k h^{n-k}]$$

for a $(k, n-k)$-power operator $f_{k,n-k}[\cdot]$, which can be considered to be defined on $X \times X$ for the same reason. Hence, if $f(x_1, \ldots, x_n) = c_n x_1 \cdots x_n$, a product of n functions in the usual sense with $c_n \in \mathcal{C}$, then we have both

$$f_n[x^n] = f_n(x, \ldots, x) = c_n x^n$$

and

$$f_{k,n-k}[x^k h^{n-k}] = f_n(x, \ldots, x, h, \ldots, h) = c_n x^k h^{n-k}$$

for any k satisfying $0 \leq k \leq n$. In particular, if $f_2 : X^2 \to Y$ is a bilinear operator, then $f_2 : (x, x) \to f_2[x^2]$ is a quadratic power operator. In the above definition, for a fixed $h \in X, f_{k,n-k}$ is a k-power operator with respect to x, and for a fixed $x \in X, f_{k,n-k}$ is an $(n-k)$-power operator with respect to h. All $(k, n-k)$-power operators are symmetric.

Throughout this chapter, we use the notation $f_n(\cdot)$ for an n-linear operator and $f_n[\cdot]$ for an n-power, respectively, to avoid any possible confusion. Moreover, we emphasize again that the notation $f_n[\cdot] : X \to Y$ will be used throughout for simplicity.

1.4. Example. The operator f_n defined in Example 1.2 is an n-power operator. Moreover, it is a $(k, n-k)$-power operator for any k satisfying $0 \leq k \leq n$.

1.5. Example. Let $f_n : x \to x^n$ be an n-power operator from X to Y, denoted $f_n[x^n]$, and $f_{k,n-k} : (x, h) \to x^k h^{n-k}$ be a $(k, n-k)$-power operator from $X \times X$ to Y, denoted $f_{k,n-k}[x^k h^{n-k}]$ as above, respectively. Then, for any constant $c \in \mathcal{C}$ and $x, h \in X$, we have the following expression:

$$f_n[(x + ch)^n]$$
$$= f_n\left[x^n + c\binom{n}{1}x^{n-1}h + \cdots + c^{n-1}\binom{n}{n-1}xh^{n-1} + c^n h^n\right]$$
$$= f_n[x^n] + c\binom{n}{1}f_{n-1,1}[x^{n-1}h] + \cdots$$
$$+ c^{n-1}\binom{n}{n-1}f_{1,n-1}[xh^{n-1}] + c^n f_n[h^n].$$

Next, let us consider the general relationship between n-linear and n-power operators.

Obviously, if f_n is an n-linear operator, then the operator g_n defined by

$$g_n : (x_1, \ldots, x_n) \rightarrow \frac{1}{n!} \sum_{1 \leq k_1, \ldots, k_n \leq n} f_n(x_{k_1}, \ldots, x_{k_n}),$$

where the sum is taken over all permutations of the indices $1, \ldots, n$, is a symmetric n-linear operator as well as an n-power operator satisfying

$$g_n(x, \ldots, x) = g_n[x^n] = f_n(x, \ldots, x).$$

This implies that different n-linear operators may determine the same n-power operator, provided that their average sums are the same on their (common) domain. In contrast, it is easily seen that an n-power operator can determine different n-linear operators. However, if we require that the n-linear operator (to be determined) be symmetric, then the result is unique. Namely, we have the following result.

1.6. Theorem. *Let $f_n : X \rightarrow Y$ be an n-power operator such that*

$$f_n[x^n] = x^n$$

for all $x \in X$. Assume it is possible to define two n-linear operators $f_n(x_1, \ldots, x_n)$ and $g_n(x_1, \ldots, x_n)$ from $f_n[x^n]$ such that

$$f_n(x_1, \ldots, x_n) = f_n[x_1, \ldots, x_n]$$

and $g_n(x, \ldots, x)$, denoted $g_n[x^n]$, is equal to x^n, namely:

$$g_n(x, \ldots, x) = g_n[x^n] = f_n[x^n],$$

where $x, x_1, \ldots, x_n \in X$. If both $f_n(x_1, \ldots, x_n)$ and $g_n(x_1, \ldots, x_n)$ are symmetric, then $f_n(x_1, \ldots, x_n) \equiv g_n(x_1, \ldots, x_n)$ for all $\mathbf{x} = (x_1, \ldots, x_n) \in X^n$.

Proof. We use mathematical induction. For $n = 1$, the result is trivial. Suppose that the result is correct up to $n - 1$. Then, it follows from 1.5 that for any $x, h \in X$ and $k = 1, \ldots, n - 1$, we have

$$f_n\left[(x + kh)^n\right] = f_n[x^n] + k\binom{n}{1}f_{n-1,1}[x^{n-1}h] + \cdots$$
$$+ k^{n-1}\binom{n}{n-1}f_{1,n-1}[xh^{n-1}] + k^n f_n[h^n].$$

Suppose, on the other hand, that

$$g_n\left[(x + kh)^n\right] = g_n[x^n] + k\binom{n}{1}g_{n-1,1}[x^{n-1}h] + \cdots$$
$$+ k^{n-1}\binom{n}{n-1}g_{1,n-1}[xh^{n-1}] + k^n g_n[h^n],$$

for some $(k, n-k)$-power operators $g_{k,n-k}[\cdot]$, $k = 1, \ldots, n-1$. Then a subtraction of the above two identities yields

$$0 = k\binom{n}{1}\left[f_{n-1,1}[x^{n-1}h] - g_{n-1,1}[x^{n-1}h]\right] + \cdots$$
$$+ k^{n-1}\binom{n}{n-1}\left[f_{1,n-1}[xh^{n-1}] - g_{1,n-1}[xh^{n-1}]\right],$$

$k = 1, \ldots, n - 1$, or

$$\begin{bmatrix} \binom{n}{1} & \binom{n}{2} & \cdots & \binom{n}{n-1} \\ 2\binom{n}{1} & 2^2\binom{n}{2} & \cdots & 2^{n-1}\binom{n}{n-1} \\ \vdots & \vdots & & \vdots \\ (n-1)\binom{n}{1} & (n-1)^2\binom{n}{2} & \cdots & (n-1)^{n-1}\binom{n}{n-1} \end{bmatrix}$$

$$\times \begin{bmatrix} f_{n-1,1}[x^{n-1}h] - g_{n-1,1}[x^{n-1}h] \\ f_{n-2,2}[x^{n-2}h^2] - g_{n-2,2}[x^{n-2}h^2] \\ \vdots \\ f_{1,n-1}[xh^{n-1}] - g_{1,n-1}[xh^{n-1}] \end{bmatrix} = 0.$$

Because the determinant of the above coefficient matrix is nonzero (see Reference [12] Vol. 2, Ch. 5, # 86), we have only zero solutions. In particular, we have

$$f_{n-1,1}[x^{n-1}h] \equiv g_{n-1,1}[x^{n-1}h].$$

Since for any $h \in X$, both $f_{n-1,1}[x^{n-1}h]$ and $g_{n-1,1}[x^{n-1}h]$ are $(n-1)$-power operators with respect to x, it follows from the induction hypothesis that

$$f_n(x_1, \ldots, x_{n-1}, h) = g_n(x_1, \ldots, x_{n-1}, h)$$

for all $x_1, \ldots, x_{n-1} \in X$ and the (arbitrary) $h \in X$. Setting $h = x_n$ completes the induction and, hence, the proof of the theorem. □

From the above theorem, we know that an n-power operator uniquely determines a symmetric n-linear operator. An immediate question is how to construct such a symmetric n-linear operator. The following theorem provides an answer.

1.7. Theorem. *Let $f_n : x \to x^n$ be an n-power operator from X to Y. Then, the following expression defines a (unique) symmetric n-linear operator from X^n to Y:*

$$f_n(x_1, \ldots, x_n) = \frac{1}{n!} \Big\{ (-1)^n f_n[x^n]$$

$$+ \sum_{k=1}^{n} (-1)^{n-k} \sum_{1 \leq j_1, \ldots, j_k \leq n} f_n[(x + x_{j_1} + \cdots + x_{j_k})^n] \Big\},$$

$$(1.1)$$

which depends on $x_1, \cdots, x_n \in X$ but is independent of $x \in X$. '

We remark that at the first glance it may seem that f_n depends on x. However, we will prove that it does not. We also emphasize that the n-linear operator f_n so defined is unique according to Theorem 1.6.

Proof. We first show that f_n is a symmetric n-linear operator independent of x. The symmetry is obvious. In order to show the independence, we introduce a "difference" operation defined inductively on any operator $f : X \rightarrow Y$ as follows:

$$\begin{cases} \Delta^1_{x_1} f(x) = f(x + x_1) - f(x), \\ \Delta^n_{x_1,\ldots,x_n} f(x) = \Delta^1_{x_n} \left(\Delta^{n-1}_{x_1,\ldots,x_{n-1}} f(x) \right), \quad n = 2, 3, \ldots . \end{cases}$$

Then, it can be verified that the right-hand side of f_n defined above in (1.1) may be rewritten as

$$\tilde{f}_n := \frac{1}{n!} \Delta^n_{x_1,\ldots,x_n} f_n[x^n].$$

To see that $\tilde{f}_n$ is independent of $x \in X$, observe from 1.5 that

$$f_n[x^n] = \sum_{k=0}^{n} \binom{n}{k} f_{n-k,k}[x^{n-k} h^k],$$

so that

$$\Delta^1_{x_1} f_n[x^n] = f_n[(x + x_1)^n] - f_n[x^n] = \sum_{k=1}^{n} \binom{n}{k} f_{n-k,k}[x^{n-k} x_1^k].$$

This implies that the highest-power term with respect to x in $\Delta^1_{x_1} f_n[x^n]$ is of degree $(n-1)$. Consequently, the highest power term with respect to x in $\Delta^1_{x_2} \left(\Delta^1_{x_1} f[x^n] \right)$ is of degree $(n-2)$. This way, we see that $\tilde{f}_n = \frac{1}{n!} \Delta^n_{x_1,\ldots,x_n} f_n[x^n]$ is independent of x.

We next show that for any $x \in X$, by setting $x_1 = \cdots = x_n = x$ we have $f_n(x,\ldots,x) = f_n[x^n]$, so that by uniqueness, $f_n(x_1,\ldots,x_n)$ is defined by $f_n[x^n]$. As a matter of fact, in this case we have $\tilde{f}_n = \frac{1}{n!} \Delta^n_{x,\ldots,x} f_n[x^n]$, and hence,

$$f_n(x,\ldots,x) = \frac{1}{n!} \Delta^n_{x,\ldots,x} f_n[x^n]$$

$$= \frac{1}{n!} \sum_{k=0}^{n} (-1)^{n-k} \binom{n}{k} f_n[(kx)^n]$$

$$= \frac{1}{n!} \sum_{k=0}^{n} (-1)^{n-k} \binom{n}{k} k^n f_n[x^n]$$

$$= \left[\frac{1}{n!} \sum_{k=0}^{n} (-1)^{n-k} \binom{n}{k} k^n \right] f_n[x^n]$$

$$= f_n[x^n],$$

where the term in large square brackets before the last equality is equal to 1 by a well-known identity (see Reference [12] Vol. 1, Ch. 1, # 41 [with $\varphi(x) = x^n$]). $\qquad\qquad\square$

We remark that Theorem 1.7 states that if an *n*-power operator is given, then a symmetric *n*-linear operator can be uniquely constructed by using Formula (1.1). Conversely, it follows from the last part of the proof that if a symmetric *n*-linear operator $f_n(\cdot, \ldots, \cdot)$ is given by Formula (1.1) using an *n*-power operator $f_n[\cdot]$, then by setting $x_1 = \cdots = x_n = x$ in $f_n(x_1, \ldots, x_n)$, we obtain $f_n(x, \ldots, x) = f_n[x^n]$.

§ 2. Bounded n-Linear Operators

In this section, we let X and Y be normed linear spaces over the field $\mathcal{C}$ of complex numbers, as defined at the beginning of Section 1, endowed respectively with norms $\|\cdot\|_X$ and $\|\cdot\|_Y$. X^n is also a normed linear space for any well-defined norm $\|\cdot\|_{X^n}$ if the linear operations between two elements of X^n are defined coordinatewise.

1.8. Definition. An n-linear operator $f_n : X^n \to Y$ is said to be *bounded* if it maps any bounded set in X^n to a bounded set in Y.

1.9. Theorem. *Let $f_n : X^n \to Y$ be an n-linear operator. Then, f_n is bounded if and only if*

$$\|f_n\| := \sup_{\substack{\|x_k\|_X = 1 \\ 1 \le k \le n}} \|f_n(x_1, \ldots, x_n)\|_Y < \infty .$$

The real number $\|f_n\|$ so defined is called the *operator norm* of the n-linear operator f_n on X^n.

Proof. Suppose that $f_n : X^n \to Y$ is bounded. Then, f_n maps the following closed bounded set

$$S = \big\{ \, (x_1, \ldots, x_n) \in X^n : \qquad \|x_k\|_X = 1, \quad k = 1, \ldots, n \, \big\}$$

to a bounded set in Y. Hence, there exists an $M > 0$ such that

$$\|f_n(x_1, \ldots, x_n)\|_Y \le M$$

for all $x_k \in S$. Taking the supremum gives $\|f_n\| \le M$.

Conversely, assume that $\|f_n\| \le M$ for some $M > 0$. Then, since f_n is a linear operator with respect to each x_k when other x_ℓ are being fixed, $\ell \neq k, k = 1, \ldots, n$, we have

$$\|f_n(x_1, \ldots, x_n)\|_Y \le \|f_n(\cdot, x_2, \ldots, x_n)\| \, \|x_1\|_X$$
$$\le \|f_n\| \, \|x_1\|_X \cdots \|x_n\|_X$$
$$\le M \|x_1\|_X \cdots \|x_n\|_X .$$

For any bounded set S in X^n, there is an $\widetilde{M}$ sufficiently large such that

$$S \subseteq \{ (x_1,\ldots,x_n) \in X^n : \quad \|x_k\|_X \leq \widetilde{M}, \quad k = 1,\ldots,n \}.$$

Hence, for any element $(x_1,\ldots,x_n) \in S$, we have

$$\|f_n(x_1,\ldots,x_n)\|_Y \leq M\widetilde{M}^n < \infty,$$

which implies that f_n maps S to a bounded set in Y. $\qquad\square$

For symmetric n-linear operators, we have stronger results.

1.10. Theorem. *Let $f_n[\cdot] : X \to Y$ be an n-power operator and $f_n : X^n \to Y$ be a symmetric n-linear operator defined via Formula (1.1) by $f_n[\cdot]$. Then, f_n is bounded if and only if $f_n[\cdot]$ is bounded in the sense that*

$$|f_n| := \sup_{\|x\|_X=1} \|f_n[x^n]\|_Y < \infty.$$

Moreover, we have

$$|f_n| \leq \|f_n\| \leq \frac{n^n}{n!}|f_n|. \tag{1.2}$$

Proof. The necessity and the first inequality of (1.2) is obvious in view of Theorem 1.9.

Suppose $|f_n| < \infty$. Recall from Formula (1.1) that

$$f_n(x_1,\ldots,x_n) = \frac{1}{n!}\Bigg\{(-1)^n f_n[x^n]$$
$$+ \sum_{k=1}^{n}(-1)^{n-k} \sum_{1\leq j_1,\ldots,j_k\leq n} f_n\big[(x + x_{j_1} + \cdots + x_{j_k})^n\big]\Bigg\}$$

in which the right-hand side is independent of $x \in X$. In particular, letting $x = -\frac{1}{2}\sum_{i=1}^n x_i$ yields

$$f_n(x_1, \ldots, x_n) = \frac{1}{n!} \left\{ (-1)^n f_n \left[\left(-\frac{1}{2} \sum_{i=1}^n x_i \right)^n \right] \right.$$

$$+ \sum_{k=1}^n (-1)^{n-k} \sum_{e_i = \pm 1} f_n \left[\left(\frac{1}{2} \sum_{i=1}^n e_i x_i \right)^n \right] \right\}$$

$$= \frac{1}{2^n n!} \left\{ f_n \left[\left(\sum_{i=1}^n x_i \right)^n \right] \right.$$

$$+ \sum_{k=1}^n (-1)^{n-k} \sum_{e_i = \pm 1} f_n \left[\left(\sum_{i=1}^n e_i x_i \right)^n \right] \right\},$$

where $\sum_{e_i = \pm 1}$ has a total of $\binom{n}{k}$ terms. It then follows that

$$\|f_n(x_1, \ldots, x_n)\|_Y$$

$$\leq \frac{1}{2^n n!} \left(1 + \sum_{k=1}^n \binom{n}{k} \right) |f_n| \left(n \cdot \max_{1 \leq k \leq n} \|x_k\|_X \right)^n$$

$$= \frac{n^n}{n!} |f_n| \left(\max_{1 \leq k \leq n} \|x_k\|_X \right)^n.$$

This proves the boundedness of f_n. Moreover, it follows that for all $x_k \in X$ with $\|x_k\|_X = 1$, $k = 1, \ldots, n$, we have

$$\|f_n\| = \sup_{\substack{\|x_k\|_X = 1 \\ 1 \leq k \leq n}} \|f_n(x_1, \ldots, x_n)\|_Y \leq \frac{n^n}{n!} |f_n|.$$

This is the second inequality of (1.2). $\qquad \Box$

We close this section by two examples that show that if an n-linear operator f_n defined in the above theorem is not symmetric then the sufficiency may not hold, and that the inequality (1.2) is sharp in the sense that both the lower and the upper bounds cannot be improved.

1.11. Example. If the n-linear operator f_n defined in Theorem 1.10 is not symmetric, then the sufficiency may not hold.

Let X be the subspace of $L_2[0, 2\pi]$ defined by

$$X = \left\{ x \in L_2[0, 2\pi] : \quad x(0) = 0, \ x \in C^1[0, 2\pi] \right\},$$

and let $Y = L_1[0, 2\pi]$. Consider the nonsymmetric bilinear operator $f_2 : X^2 \to Y$ defined by

$$f_2(x_1, x_2) = \int_0^t \frac{d}{dt}[x_1(t)]x_2(t)dt, \quad x_1, x_2 \in X, \quad t \in [0, 2\pi].$$

The corresponding quadratic power operator is given by

$$f_2[x^2] = \frac{1}{2} \int_0^t \frac{d}{dt}\left[x^2(t)\right]dt = \frac{1}{2}x^2(t),$$

so that $|f_2| = \frac{1}{2}$. We will show, however, that as a bilinear operator f_2 is unbounded.

Let $h_k(t) = \frac{1}{\sqrt{\pi}}\sin(kt), k = 1, 2, \dots$. Then $h_k \in X$ with $\|h_k\|_X = 1$ for all $k = 1, 2, \dots$. Define a sequence $\{\tilde{h}_k\}$ in X by

$$\tilde{h}_k(t) = \begin{cases} 0 & t = 0 \\ \text{positive} & 0 < t < \dfrac{1}{k} \\ \dfrac{1}{\sqrt{\pi}}\cos(kt) & \dfrac{1}{k} \leq t \leq 2\pi, \end{cases}$$

where "positive" means that it can be arbitrarily defined as long as it has positive values and $\tilde{h}_k \in X$.

Since both $h_k(t)$ and $\tilde{h}_k(t)$ are non-negative on $[0, \frac{1}{k}]$, we have

$$f_2(h_k, \tilde{h}_k) = \int_0^t \dot{h}_k(t)\tilde{h}_k(t)dt$$

$$= \int_0^t \frac{k}{\sqrt{\pi}}\cos(kt)\tilde{h}_k(t)dt$$

$$\geq \int_{1/k}^t \frac{k}{\pi}\cos^2(kt)dt$$

$$= \frac{k}{2\pi}\left(t + \frac{\sin(2kt)}{2k} - \frac{1}{k} - \frac{\sin(2)}{2k}\right).$$

Consequently, we have

$$
\|f_2(h_k, \tilde{h}_k)\|_Y = \int_0^{1/k} |f_2(h_k, \tilde{h}_k)| dt + \int_{1/k}^{2\pi} |f_2(h_k, \tilde{h}_k)| dt
$$

$$
\geq \int_{1/k}^{2\pi} \frac{k}{2\pi} \left| t + \frac{\sin(2kt)}{2k} - \frac{1}{k} - \frac{\sin(2)}{2k} \right| dt
$$

$$
\to \infty \ (k \to \infty).
$$

This implies that f_2 is unbounded.

1.12. Example. Both the lower and upper bounds of Inequality
(1.2) cannot be improved.

Let $X = R^2, Y = R^1$, and $X^2 = R^2 \times R^2$. Consider the
bilinear functional f_2 defined by

$$
f_2(x, h) = x_1 h_1 - x_2 h_2,
$$

where $x = \begin{bmatrix} x_1 \\ x_2 \end{bmatrix}$ and $h = \begin{bmatrix} h_1 \\ h_2 \end{bmatrix}$.

First, let X be endowed with the Euclidean norm: $\|x\|_X = \sqrt{x_1^2 + x_2^2}$. Then, by the Schwartz inequality we have

$$
\|f_2(x, h)\|_Y = |x_1 h_1 - x_2 h_2| \leq \|x\|_X \|h\|_X .
$$

Hence, $\|f_2\| \leq 1$. Let $x = \begin{bmatrix} 1 \\ 0 \end{bmatrix}$. Then, we have

$$
\|f_2[x^2]\|_Y = |x_1^2 - x_2^2| = 1 = \|x\|_X ,
$$

so that $|f_2| \geq 1$. It follows from the left-hand inequality of (1.2)
that

$$
1 \leq |f_2| \leq \|f_2\| \leq 1,
$$

that is, $|f_2| = \|f_2\|$. This implies that the first inequality of (1.2)
cannot be improved.

Secondly, let X be endowed with the ℓ_∞-norm:

$$
\|x\|_X = \max\{|x_1|, |x_2|\}.
$$

On the one hand, we have, for any $x \in X$ with $\|x\|_X = 1$, that

$$
\|f_2[x^2]\|_Y = |x_1^2 - x_2^2| \leq |x_1^2| \leq \|x\|_X ,
$$

so that $|f_2| \leq 1$. On the other hand, let $x = \begin{bmatrix} 1 \\ 1 \end{bmatrix}$ and $h = \begin{bmatrix} 1 \\ -1 \end{bmatrix}$. Then, we have $\|x\|_X = \|h\|_X = 1$ and

$$\|f_2(x, h)\|_Y = |x_1 h_1 - x_2 h_2| = 2 ,$$

so that $\|f_2\| = 2$. It then follows from the right-hand inequality of (1.2) that

$$2 = \|f_2\| \leq \frac{2^2}{2!} |f_2| \leq 2 .$$

Hence, $\|f_2\| = 2|f_2|$. This implies that the right-hand inequality of (1.2) cannot be improved.

§ 3. Normed Linear Spaces of Bounded n-Linear Operators

Let X and Y be normed linear spaces and let $X^n = X \times \cdots \times X$ as before. Consider the family of bounded n-linear operators from X^n to Y and the family of bounded symmetric n-linear operators from X^n to Y. It can be easily seen that they are both linear spaces. We will further show that they are also normed linear spaces when equipped with the operator norm

$$\|f_n\| := \sup_{\substack{\|x_k\|_X = 1 \\ 1 \le k \le n}} \|f_n(x_1, \ldots, x_n)\|_Y ,$$

defined as in 1.9. With the notation

$$\mathcal{L}(X^n, Y) = \{\, f_n : X^n \to Y : \quad \|f_n\| < \infty \,\} \tag{1.3}$$

and

$$\mathcal{L}^s(X^n, Y) = \{f_n : X^n \to Y : f_n \text{ symmetric}, \|f_n\| < \infty\}, \tag{1.4}$$

we first establish the following result for $\mathcal{L}(X^n, Y)$.

1.13. Theorem. $\mathcal{L}(X^n, Y)$ *is a normed linear space. Moreover, for any* $k : 1 \le k \le n - 1$, *there is an isomorphism* T_k *such that*

$$T_k\{\mathcal{L}(X^n, Y)\} = \mathcal{L}(X^k, \mathcal{L}(X^{n-k}, Y)).$$

Finally, if Y *is complete then so is* $\mathcal{L}(X^n, Y)$.

We remark that the theorem implies that if X is a normed linear space and Y is a Banach space, then $\mathcal{L}(X^n, Y)$ is also a Banach space (of bounded n-linear operators) for each $n, n = 1, 2, \ldots$.

Proof. We use mathematical induction. For $n = 1, \mathcal{L}(X, Y)$ is a standard normed linear space of bounded linear operators. Suppose that $\mathcal{L}(X^{n-k}, Y)$ is a normed linear space for $1 \le k \le n - 1$. We then verify the result for $k = 0$. For this purpose, we first establish a one-to-one correspondence between $\mathcal{L}(X^n, Y)$ and $\mathcal{L}(X^k, \mathcal{L}(X^{n-k}, Y))$, $1 \leq k \leq n-1$, as follows.

For any $f_n \in \mathcal{L}(X^n, Y)$ and $1 \leq k \leq n-1$, fix $x_1, \ldots, x_k \in X$. Then f_n is a bounded $(n-k)$-linear operator with respect to $x_{k+1}, \ldots, x_n \in X$. Hence, there exists an $\tilde{f}_{n-k} \in \mathcal{L}(X^{n-k}, Y)$ with parameters $x_1, \ldots, x_k$, namely, $\tilde{f}_{n-k} = \tilde{f}_{n-k}(x_1, \ldots, x_k)$, such that

$$\tilde{f}_{n-k} : (x_1, \ldots, x_n) \to \tilde{f}_{n-k}(x_1, \ldots, x_k)(x_{k+1}, \ldots, x_n), \qquad (1.5)$$

with

$$\tilde{f}_{n-k}(x_1, \ldots, x_k)(x_{k+1}, \ldots, x_n) = f_n(x_1, \ldots, x_n) \in Y.$$

However, $\tilde{f}_{n-k}$ is also a bounded k-linear operator with respect to $x_1, \ldots, x_k \in X$, so that we can define an operator norm $\| \cdot \|$ for $\tilde{f}_{n-k}$ via

$$\|\tilde{f}_{n-k}\| := \sup_{\substack{\|x_i\|_X = 1 \\ 1 \leq i \leq k}} \|\tilde{f}_{n-k}(x_1, \ldots, x_k)\|_{\mathcal{L}(X^{n-k}, Y)}.$$

Consequently, we have

$$
\begin{aligned}
\|\tilde{f}_{n-k}\| &= \sup_{\substack{\|x_i\|_X = 1 \\ 1 \leq i \leq k}} \|\tilde{f}_{n-k}(x_1, \ldots, x_k)\|_{\mathcal{L}(X^{n-k}, Y)} \\
&= \sup_{\substack{\|x_i\|_X = 1 \\ 1 \leq i \leq k}} \sup_{\substack{\|x_j\|_X = 1 \\ k+1 \leq j \leq n}} \|\tilde{f}_{n-k}(x_1, \ldots, x_k)(x_{k+1}, \ldots, x_n)\|_Y \\
&= \sup_{\substack{\|x_k\|_X = 1 \\ 1 \leq k \leq n}} \|f_n(x_1, \ldots, x_n)\|_Y \\
&= \|f_n\|.
\end{aligned}
$$

Thus, we have established a one-to-one and norm-preserving correspondence between $\mathcal{L}(X^n, Y)$ and $\mathcal{L}(X^k, \mathcal{L}(X^{n-k}, Y))$, $1 \leq k \leq n-1$. Hence, for any $1 \leq k \leq n-1$, there is an isomorphism T_k such that

$$T_k\{\mathcal{L}(X^n, Y)\} = \mathcal{L}(X^k, \mathcal{L}(X^{n-k}, Y)).$$

It follows from the induction hypothesis that $\mathcal{L}(X^n, Y)$ is a normed linear space. Finally, the completeness of Y implies the completeness of $\mathcal{L}(X^k, \mathcal{L}(X^{n-k}, Y))$ and hence that of $\mathcal{L}(X^n, Y)$ since T_k is an isomorphism. $\qquad \square$

As to the linear space $\mathcal{L}^s(X^n, Y)$ defined in (1.4), we have a similar result, shown in the following theorem.

1.14. Theorem. $\mathcal{L}^s(X^n, Y)$ *is a normed linear space. Moreover, for any $k : 1 \leq k \leq n-1$, there is an isomorphism T_k such that*

$$T_k\{\mathcal{L}^s(X^n, Y)\} \subset \mathcal{L}^s\left(X^k, \mathcal{L}^s(X^{n-k}, Y)\right).$$

Finally, if Y is complete then so is $\mathcal{L}^s(X^n, Y)$.

The proof is also similar. The only difference is that $\mathcal{L}^s(X^n, Y)$ is isomorphic to a subspace of $\mathcal{L}^s\left(X^k, \mathcal{L}^s(X^{n-k}, Y)\right)$. This is because the symmetry of f_n implies the symmetries of both $\tilde{f}_{n-k}$ and $\tilde{f}_{n-k}(x_1, \ldots, x_k)$, but the converse is not true in general.

The following result is a natural generalization of the well-known uniform boundedness theorem from a family of bounded linear operators to a family of bounded n-linear operators.

1.15. Theorem. *Let X be a Banach space and Y be a normed linear space. Let $\mathcal{L}(X^n, Y)$ be the family of bounded n-linear operators defined in (1.3). If for any fixed $(x_1, \ldots, x_n) \in X^n$ we have*

$$\sup_{f_n \in \mathcal{L}(X^n, Y)} \|f_n(x_1, \ldots, x_n)\|_Y < \infty,$$

then

$$\sup_{f_n \in \mathcal{L}(X^n, Y)} \|f_n\| < \infty,$$

where $\|f_n\|$ is the operator norm of f_n defined in Theorem 1.9.

Proof. We use mathematical induction. For $n = 1$, this is the standard uniform boundedness theorem. Fix $x_1 \in X$. By (1.5), there exists an $\tilde{f}_{n-1} \in \mathcal{L}(X^{n-1}, Y)$ such that

$$f_n(x_1, \ldots, x_n) = \tilde{f}_{n-1}(x_1)(x_2, \ldots, x_n),$$

and then by the assumption, for any fixed $(x_2, \ldots, x_n) \in X^{n-1}$, we have

$$\sup_{\tilde{f}_{n-1} \in \mathcal{L}(X^{n-1}, Y)} \|\tilde{f}_{n-1}(x_1)(x_2, \ldots, x_n)\|_Y$$

$$\leq \sup_{f_n \in \mathcal{L}(X^n, Y)} \|f_n(x_1, \ldots, x_n)\|_Y < \infty.$$

It follows from the induction hypothesis that

$$\sup_{\tilde{f}_{n-1}\in\mathcal{L}(X^{n-1},Y)} \|\tilde{f}_{n-1}(x_1)\| < \infty.$$

Viewing $\tilde{f}_{n-1}(x_1)$ as a bounded linear operator with respect to $x_1 \in X$ for any fixed $(x_2, \ldots, x_n) \in X^{n-1}$, as we did in the proof of Theorem 1.13 with $k = 1$ therein, we can apply the standard uniform boundedness theorem to the linear operator $\tilde{f}_{n-1}$ to conclude that

$$\sup_{f_{n-1}\in\mathcal{L}(X^{n-1},Y)} \|\tilde{f}_{n-1}\| < \infty.$$

It follows again from the proof of Theorem 1.13 that

$$\sup_{f_n\in\mathcal{L}(X^n,Y)} \|f_n\| < \infty.$$

This completes the mathematical induction. $\qquad\square$

§ 4. Continuous n-Linear Operators

In Section 2, we have studied in some detail both bounded n-linear operators and symmetric bounded n-linear operators. As is well known, unlike the linear case, continuity and boundedness are not equivalent in nonlinear settings. One simple example is the unit step function, which is bounded but not continuous. On the other hand, there are many examples of continuous but unbounded nonlinear operators. The following example shows that a continuous nonlinear functional may be unbounded on a ball, which is a bounded, closed, and convex subset, in a Banach space.

1.16. Example. Let $X = \ell_1$ and consider the functional

$$f(\mathbf{x}) = \sum_{k=1}^{\infty} k(|x_k| - 1)_+$$

where $\mathbf{x} = (x_1, x_2, \dots) \in \ell_1$ and

$$(|x_k| - 1)_+ = \begin{cases} |x_k| - 1 & \text{if } |x_k| - 1 \geq 0 \\ 0 & \text{otherwise}. \end{cases}$$

Since $x_k \to 0$ as $k \to \infty$, the sum in the definition of $f(\mathbf{x})$ has only finitely many nonzero terms. Hence, for any $\mathbf{x}, \tilde{\mathbf{x}} \in \ell_1$,

$$|f(\mathbf{x}) - f(\tilde{\mathbf{x}})| \leq \sum_{k=1}^{N} k|(|x_k| - 1) - (|\tilde{x}_k| - 1)|$$

$$\leq N\|\mathbf{x} - \tilde{\mathbf{x}}\|_{\ell_1}$$

for some integer N, depending on both $\mathbf{x}$ and $\tilde{\mathbf{x}}$. This implies that $f(\mathbf{x})$ is continuous (but not uniformly) on ℓ_1.

For any fixed $\epsilon > 0$, define a sequence $\{\mathbf{x}^{(k)}\}$ of elements of ℓ_1, where $\mathbf{x}^{(k)} = (x_1^{(k)}, x_2^{(k)}, \dots)$ with

$$x_i^{(k)} = \begin{cases} 1 + \epsilon & \text{if } i = k \\ 0 & \text{if } i \neq k. \end{cases}$$

Then, $\mathbf{x}^{(k)} \in B = \{\mathbf{x} \in \ell_1 : \|\mathbf{x}\|_{\ell_1} \leq 1 + \epsilon\}$ for all $k = 1, 2, \dots$. However, $f(\mathbf{x}^{(k)}) = \epsilon k \to \infty$ as $k \to \infty$. This implies that f is unbounded on the ball B in ℓ_1.

Therefore, it is necessary to consider the relationship between the continuity and boundedness for a bounded n-linear operator and for a symmetric bounded n-linear operator. Fortunately, a bounded n-linear operator is "not very nonlinear" in the sense that it behaves almost like a bounded linear operator. More precisely, we have the following result.

1.17. Theorem. *Let $f_n : X^n \to Y$ be an n-linear operator. Then f_n is bounded if and only if $f_n : X^n \to Y$ is continuous in the sense that it is continuous in the n-tuple $\mathbf{x} = (x_1, \ldots, x_n) \in X^n$. If, furthermore, X is a Banach space, then f_n is bounded if and only if f_n is continuous in each $x_k, k = 1, \ldots, n$.*

Proof. Suppose that $f_n : X^n \to Y$ is bounded. Then, for any $\delta > 0$, conditions $\|x_k\|_X \le \delta$ and $\|\tilde{x}_k\|_X \le \delta$ for all $k = 1, \ldots, n$, imply

$$\|f_n(x_1, \ldots, x_n) - f_n(\tilde{x}_1, \ldots, \tilde{x}_n)\|_Y$$

$$\le \sum_{k=1}^{n} \|f_n(x_1, \ldots, x_{k-1}, x_k - \tilde{x}_k, \tilde{x}_{k+1}, \ldots, \tilde{x}_n)\|_Y$$

$$\le \|f_n\|\delta^{n-1} \sum_{k=1}^{n} \|x_k - \tilde{x}_k\|_X ,$$

where $\|f_n\|$ is the operator norm of f_n defined in Theorem 1.9. This implies that $f_n : X^n \to Y$ is continuous in the n-tuple $\mathbf{x} = (x_1, \ldots, x_n) \in X^n$.

Conversely, suppose that $f_n : X^n \to Y$ is continuous in the n-tuple $\mathbf{x} = (x_1, \ldots, x_n)$ but is unbounded. Then, there exists a sequence $\{(x_1^{(k)}, \ldots, x_n^{(k)}) : k = 1, 2, \cdots\}$ in X^n with $\|x_i^{(k)}\|_X = 1, i = 1, \ldots, n$, for all k, such that

$$c_k := \|f_n(x_1^{(k)}, \ldots, x_n^{(k)})\|_Y \to \infty \qquad \text{as} \quad k \to \infty .$$

Let
$$\tilde{x}_i^{(k)} = c_k^{-1/(n+1)} x_i^{(k)}, \quad i = 1, \ldots, n; k = 1, 2, \ldots .$$

Then, $\tilde{x}_i^{(k)} \to 0$ as $k \to \infty$ for each $i, i = 1, \ldots, n$. However,

$$\|f_n(\tilde{x}_1^{(k)}, \ldots, \tilde{x}_n^{(k)})\|_Y = c_k^{-n/(n+1)} \|f_n(x_1^{(k)}, \ldots, x_n^{(k)})\|_Y$$

$$= c_k^{-n/(n+1)} c_k \to \infty \qquad \text{as} \quad k \to \infty .$$

This implies that f_n is not continuous at the origin $\mathbf{0} = (0, \ldots, 0)$, both in the n-tuple $\mathbf{x}$ and in each $x_k, k = 1, \ldots, n$, which is a contradiction. Hence, $f_n : X^n \to Y$ must be bounded. This completes the proof of the first part of the theorem.

Next, suppose that X is a Banach space. We only need to show that if f_n is continuous in each $x_k, k = 1, \ldots, n$, then f_n is bounded. Another direction has actually been verified above.

Again, we use mathematical induction. For $n = 1$, the result has been proved above. Consider the case $n = 2$. By assumption, f_n is continuous in x_1 and x_2, respectively. Since f_n is linear with respect to either x_1 or x_2, it is bounded in x_1 and x_2. Hence, there exist $\tilde{f}_1(x_1) \in \mathcal{L}(X, Y)$ and $\hat{f}_1(x_2) \in \mathcal{L}(X, Y)$ such that

$$f_2(x_1, x_2) = \tilde{f}_1(x_1)x_2 = \hat{f}_1(x_2)x_1$$

(see (1.5)). For any $x_1 \in X$ with $\|x_1\|_X = 1$, we have

$$\|\tilde{f}_1(x_1)x_2\|_Y = \|\hat{f}_1(x_2)x_1\|_Y \leq \|\hat{f}_1(x_2)\|\,\|x_1\|_X = \|\hat{f}_1(x_2)\|,$$

where $\|\cdot\|$ is the operator norm defined in Theorem 1.9. This implies that for a fixed $x_2 \in X, \|\tilde{f}_1(x_1)x_2\|_Y < \infty$, so that by the standard uniform boundedness theorem there is a constant M, independent of x_1, such that $\|\tilde{f}_1(x_1)\| \leq M$ for all $x_1 \in X$ with $\|x_1\|_X = 1$. Consequently, for any $x_1 \in X$ with $\|x_1\|_X = 1$ and for any $x_2 \in X$, we have

$$\|f_2(x_1, x_2)\|_Y = \|\tilde{f}_1(x_1)x_2\|_Y \leq M\|x_2\|_X,$$

so that for any $x_1, x_2 \in X$ with $x_1 \neq 0$,

$$\|f_2(x_1, x_2)\|_Y = \left\|f_2\left(\frac{x_1}{\|x_1\|_X}, x_2\right)\right\|_Y \|x_1\|_X \leq M\|x_1\|_X\|x_2\|_X.$$

This implies that f_2 is bounded. Applying induction to cases $n > 2$, where Theorems 1.13 and 1.15 may be used, establishes the boundedness of f_n. $\qquad\square$

Since if a symmetric n-linear operator is continuous in one $x_k, k = 1, \ldots, n$, then it is also continuous in each $x_i, i = 1, \ldots, n$, the following result is immediate.

1.18. Corollary. *Let X be a Banach space and $f_n : X \to Y$ be a symmetric n-linear operator. If f_n is continuous in some $x_k, k = 1, \ldots, n$, then f_n is bounded.*

The last result of this section stated below ties in the boundedness of a symmetric n-linear operator with the continuity of a corresponding n-power operator. In order to be able to establish this elegant result, we first need the following lemma.

1.19. Lemma. *Let $f_n[\cdot] : X \to Y$ be an n-power operator. If for an $x_0 \in X$,*

$$\sup_{\|x - x_0\|_X \leq \delta} \|f_n[x^n]\|_Y \leq M$$

for some positive numbers δ and M, then

$$\sup_{\|x\|_X \leq \delta} \|f_n[x^n]\|_Y \leq 2^{n-1} M.$$

Proof. Let $x = x_0 + \lambda h$ with $\|h\|_X \leq \delta$ and $0 \leq \lambda \leq 1$. Then, we have

$$f_n[x^n] = f_n[(x_0 + \lambda h)^n] = \sum_{k=0}^{n} \lambda^k \binom{n}{k} f_{n-k,k}[x_0^{n-k} h^k],$$

(see Example 1.5). If $\|f_n[h^n]\|_Y = 0$, then obviously we have

$$0 = \|f_n[h^n]\|_Y \leq 2^{n-1} M.$$

Suppose that $\|f_n[h^n]\|_Y \neq 0$. It follows from the Hahn-Banach theorem that there exists a $y^* \in Y^*$, the dual space of Y, with $\|y^*\| = 1$, such that

$$y^*(f_n[h^n]) = \|f_n[h^n]\|_Y.$$

Consequently, we have

$$y^*(f_n[x^n]) = y^*\big(f_n[(x_0 + \lambda h)^n]\big)$$

$$= \sum_{k=0}^{n} \lambda^k \binom{n}{k} y^*\big(f_{n-k,k}[x_0^{n-k} h^k]\big)$$

$$= \|f_n[h^n]\|_Y \big(\lambda^n + a_{n-1}\lambda^{n-1} + \cdots + a_1\lambda + a_0\big)$$

where

$$a_k = \binom{n}{k} \frac{y^*(f_{n-k,k}[x_0^{n-k}h^k])}{\|f_n[h^n]\|_Y}, \quad k = 0, 1, \ldots, n-1.$$

Since

$$\max_{0 \leq \lambda \leq 1} |\lambda^n + a_{n-1}\lambda^{n-1} + \cdots + a_1\lambda + a_0| \geq \frac{1}{2^{n-1}}$$

(see Reference [12] Vol. 2, Ch. 6, # 62), it follows that

$$\|f_n[h^n]\|_Y \leq 2^{n-1} \sup_{x=x_0+\lambda h} |y^*(f_n[x^n])|$$

$$\leq 2^{n-1}\|y^*\| \sup_{0 \leq \lambda \leq 1} \|f_n[(x_0 + \lambda h)^n]\|_Y$$

$$\leq 2^{n-1}M.$$

Hence, we have

$$\|f_n[h^n]\|_Y \leq 2^{n-1}M$$

for all $h \in X$ with $\|h\|_X \leq \delta$. Replacing h by an arbitrary element x satisfying $\|x\|_X \leq \delta$ completes the proof of the lemma. □

Based on this result, we can now establish the following theorem.

1.20. Theorem. *Let $f[\cdot] : X \to Y$ be an n-power operator and let $f_n : X^n \to Y$ be a symmetric n-linear operator generated by $f_n[\cdot]$. If $f[\cdot]$ is continuous at some $x_0 \in X$, then f_n is bounded.*

Proof. Since $f_n[\cdot]$ is continuous at $x_0 \in X$, there exists a $\delta > 0$ such that $\|x - x_0\|_X \leq \delta$ implies $\|f_n[x^n] - f_n[x_0^n]\|_Y \leq 1$. Hence,

$$\|f_n[x^n]\|_Y \leq \|f_n[x_0^n]\|_Y + \|f_n[x^n] - f_n[x_0^n]\|_Y$$

$$\leq \|f_n[x_0^n]\|_Y + 1.$$

Let $M = \|f_n[x_0^n]\|_Y + 1$. It then follows from Lemma 1.19 that

$$\sup_{\|x\|_X \leq \delta} \|f_n[x^n]\|_Y \leq 2^{n-1}M,$$

or

$$\sup_{\|x\|_X = 1} \|f_n[x^n]\|_Y \leq \frac{2^{n-1}}{\delta^n}M,$$

so that by Theorem 1.10, f_n is bounded. □

§ 5. Differentiation of Nonlinear Operators

The theory of n-linear operators developed in the previous sections depends only on the simple analytic notion of boundedness and continuity of the operators. The nonlinear analytic mappings to be considered later in this chapter will require that the operator be, in addition, differentiable in a certain generalized sense. This section is devoted to an extension of the basic idea of differentiation from calculus to nonlinear operators on normed linear spaces.

1.21. Definitions. Let X and Y be normed linear spaces and Ω be an open set in X. An operator $T : \Omega \to Y$ is said to be *differentiable* at a point $x_0 \in \Omega$ in the sense of Fréchet if there exists a bounded linear operator $L(x_0) : X \to Y$, depending on x_0, such that for $h \in X$ and $x_0 + h \in \Omega$,

$$\lim_{\|h\|_X \to 0} \frac{\|T(x_0 + h) - T(x_0) - L(x_0)(h)\|_Y}{\|h\|_X} = 0 \,.$$

In this case, $L(x_0) := L(x_0)(\cdot)$ is called the *first Fréchet derivative* of T at x_0, denoted $L(x_0) = T'(x_0)$, and $dT(x_0, h) := L(x_0)(h)$ is called the *Fréchet differential* of T at x_0. Moreover, if $T : \Omega \to Y$ is differentiable at every point $x \in \Omega$, then $L(\cdot) = T'(\cdot)$ is a mapping from Ω to $\mathcal{L}(X, Y)$, the family of bounded linear operators defined on X with values in Y, and is called the *derivative mapping* of T.

Here, a derivative mapping $L(\cdot) : \Omega \to \mathcal{L}(X, Y)$ is understood in the following way: For any fixed $x_0 \in \Omega$, $L(x_0)(\cdot)$ is a bounded linear operator from X to Y. Note, however, that if we let x_0 vary in Ω, then the operator $L(x_0)(\cdot)$, considering the varying $x_0 \in \Omega$ as a variable, is not linear in general, as can be seen from Example 1.23 given below.

We remark that the names *derivative* and *differential* will be used later to denote the Fréchet derivative and Fréchet differential, respectively, unless otherwise noted. We also remark that although the domain of T is Ω (or a subset containing Ω), the domain of $L(x_0) = T'(x_0)$ is the entire space X for any $x_0 \in \Omega$. Moreover, although both T and the derivative mapping $L = T'$ have the common domain Ω, their ranges are in entirely different spaces. Of course, for each fixed $x_0 \in \Omega$, T and its derivative $L(x_0) = T'(x_0)$ have their ranges in the same space, Y.

The following result can be easily verified:

1.22. Theorem. *Let X, Y, Z be normed linear spaces and $\Omega \subseteq X$ an open set.*

(1) *Fréchet differentiation is a linear operation, namely: If $T_1, T_2 : \Omega \to Y$ are both differentiable at $x_0 \in \Omega$, then for any $a, b \in \mathcal{C}$,*

$$(aT_1 + bT_2)'(x_0) = aT_1'(x_0) + bT_2'(x_0).$$

(2) *If $L : X \to Y$ is a bounded linear operator, then its derivative at any point $x_0 \in X$ is itself, namely, $L'(x_0)(x) = L(x)$ for any fixed $x_0 \in X$ and for all $x \in X$.*

(3) *If $T : X \to Y$ is a constant operator in the sense that $T(x) = y_0$ for a fixed $y_0 \in Y$ for all $x \in X$, then $T'(x) = 0$ for all $x \in X$.*

(4) *Denote by $\mathcal{R}(T)$ the range of the operator T. Let $T_1 : \Omega \to Y$ and $T_2 : \mathcal{R}(T_1) \to Z$ be two differentiable operators at $x_0 \in \Omega$ and at $T_1(x_0) \in \mathcal{R}(T_1)$, respectively. Then, the composite operator $T_2 T_1 : \Omega \to Z$ defined by $T_2 T_1(x) = T_2\big(T_1(x)\big)$ is differentiable at $x_0 \in \Omega$ with*

$$(T_2 T_1)'(x_0) = T_2'\big(T_1(x_0)\big) T_1'(x_0).$$

In particular, if $T_2 = L$ is a bounded linear operator from Ω to Y, then

$$(L T_1)'(x_0) = L T_1'(x_0).$$

We remark that the last statement in the theorem implies that, in the differential calculations on normed linear spaces, a bounded linear operator plays a role similar to a constant multiplier in the ordinary calculus.

In order to actually find the derivative L for a nonlinear operator $T : \Omega \to Y$, we may first write the difference $T(x_0 + h) - T(x_0)$ in the form

$$T(x_0 + h) - T(x_0) = L(x_0, h)(h) + \epsilon(x_0, h),$$

where $L(x_0, h)$ is a bounded linear operator for any given x_0 and h such that

$$\lim_{\|h\|_X \to 0} L(x_0, h) = L(x_0)$$

for a bounded linear operator $L(x_0) : X \to Y$, and

$$\lim_{\|h\|_X \to 0} \frac{\|\epsilon(x_0, h)\|_Y}{\|h\|_X} = 0 \,.$$

If these two limits exist, then we have

$$\lim_{\|h\|_X \to 0} L(x_0, h) = T'(x_0) = L(x_0) \,.$$

Moreover, if $L(x_0, h)$, considered as a linear operator from X to $\mathcal{L}(X, Y)$, is continuous in h for any h in a ball $B_r(0)$ of center 0 and radius $r > 0$, then

$$L(x_0, 0) = T'(x_0) = L(x_0) \,.$$

1.23. Example. Let $C[0, 1]$ be the family of continuous functions defined on $[0,1]$ and $T : C[0, 1] \to C[0, 1]$ be a nonlinear integral operator defined by

$$T(x) := T(x)(t) = x(t) \int_0^1 K(t, \tau) x(\tau) d\tau \,, \quad 0 \le t \le 1 \,,$$

where $K \in L_1\big([0, 1] \times [0, 1]\big)$. Then, we have

$$T(x_0 + h) - T(x_0)$$
$$= x_0(t) \int_0^1 K(t, \tau) h(\tau) d\tau + h(t) \int_0^1 K(t, \tau) x_0(\tau) d\tau$$
$$+ h(t) \int_0^1 K(t, \tau) h(\tau) d\tau$$

and

$$L(x_0, h)(\ ) = x_0(t) \int_0^1 K(t, \tau)(\ ) d\tau + (\ ) \int_0^1 K(t, \tau) x_0(\tau) d\tau$$
$$+ (\ ) \int_0^1 K(t, \tau) h(\tau) d\tau \,;$$

or

$$L(x_0, h)(\ ) = x_0(t) \int_0^1 K(t,\tau)(\ )d\tau + (\ ) \int_0^1 K(t,\tau)x_0(\tau)d\tau$$

$$+ h(t) \int_0^1 K(t,\tau)(\ )d\tau \,.$$

It is clear that for any given x_0 and h in $C[0,1]$, both of the above two $L(x_0, h)$ define a bounded linear operator from $C[0,1]$ to $C[0,1]$, which is a linear operator continuous in h. It then follows from the above limiting procedure that

$$L(x_0)(\ ) = T'(x_0)(\ ) = L(x_0, 0)(\ )$$

$$= x_0(t) \int_0^1 K(t,\tau)(\ )d\tau + (\ ) \int_0^1 K(t,\tau)x_0(\tau)d\tau \,.$$

For any $\tilde{x} \in C[0,1]$, we have

$$L(x_0)(\tilde{x}) = x_0(t) \int_0^1 K(t,\tau)\tilde{x}(\tau)d\tau + \tilde{x}(t) \int_0^1 K(t,\tau)x_0(\tau)d\tau \,.$$

Here, $L(x_0)(\cdot)$ is linear in $\tilde{x}$ for any fixed x_0, but is nonlinear if considering both x_0 and $\tilde{x}$ as variables.

1.24. Example. Let $\Delta : C^2\big([0,1] \times [0,1]\big) \to C\big([0,1] \times [0,1]\big)$ be the Laplacian partial differential operator and consider a nonlinear operator $T : C^2\big([0,1] \times [0,1]\big) \to C\big([0,1] \times [0,1]\big)$ defined by

$$T(w) = \Delta w - w^2 \,.$$

It can be verified, by following the above procedure, that

$$L(w_0)(\ ) = T'(w_0)(\ ) = \Delta(\ ) - 2w_0(\ ) \,.$$

Since Δ is a linear operator, for any $\tilde{w} \in C^2\big([0,1] \times [0,1]\big)$, we have

$$L(w_0)(\tilde{w}) = \Delta\tilde{w} - 2w_0\tilde{w} \,.$$

Here, again, $L(w_0)(\tilde{w})$ is linear in $\tilde{w}$ for any fixed w_0, but is nonlinear if considering both w_0 and $\tilde{w}$ as variables.

The following result provides a necessary and sufficient condition for the differentiability of a nonlinear operator $T : \Omega \to Y$ at a point $x_0 \in \Omega$.

1.25. Theorem. *A nonlinear operator* $T : \Omega \to Y$ *is differentiable at a point* $x_0 \in \Omega$ *if and only if* T *is continuous at* x_0 *and there exists a bounded linear operator* $L \in \mathcal{L}(X, Y)$ *such that*

$$\lim_{\|h\|_X \to 0} \frac{\|T(x_0 + h) - T(x_0) - L(x_0)(h)\|_Y}{\|h\|_X} = 0 \qquad (1.6)$$

for all $h \in X$ *satisfying* $x_0 + h \in \Omega$.

Proof. By Definition 1.21, for necessity we only have to show that if T is differentiable at x_0 then T is continuous at x_0. This is evident because if T is differentiable at x_0, we have $T'(x_0) = L(x_0)$ bounded (in the operator norm), so that

$$\|T(x_0 + h) - T(x_0)\|_Y$$
$$\leq \|T(x_0 + h) - T(x_0) - L(x_0)(h)\|_Y + \|L(x_0)\| \, \|h\|_X \to 0$$

as $\|h\|_X \to 0$. This implies that T is continuous at x_0.

Again by Definition 1.21, for sufficiency we only have to show that if a linear operator $L(x_0)$ satisfies (1.6) then it is bounded. As a matter of fact, since T is continuous at x_0 and since (1.6) holds for a linear operator $L : X \to Y$, for any $\epsilon > 0$ there exists a $\delta > 0$ such that whenever $\|h\|_X \leq \delta$, we have both

$$\|T(x_0 + h) - T(x_0)\|_Y \leq \epsilon$$

and

$$\|T(x_0 + h) - T(x_0) - L(x_0)(h)\|_Y \leq \epsilon .$$

Consequently, we have

$$\|L(x_0)h\|_Y \leq \|T(x_0 + h) - T(x_0) - L(x_0)(h)\|_Y$$
$$+ \|T(x_0 + h) - T(x_0)\|_Y \leq 2\epsilon ,$$

so that

$$\|L(x_0)\| = \sup_{\|h\|_X = 1} \|L(x_0)(h)\|_Y \leq \frac{2\epsilon}{\delta} .$$

This implies that $L(x_0) : X \to Y$ is bounded and hence $L(x_0) \in \mathcal{L}(X, Y)$, completing the proof of the theorem. $\qquad\qquad \square$

§ 6. Higher-Order Derivatives of Nonlinear Operators

Based on the theory developed in the previous section, we now consider higher-order derivatives of nonlinear operators, which will be needed in the Taylor series expansion of a nonlinear operator discussed later in this section. Since we are not concerned with higher-order differentials of nonlinear operators, we will not discuss them in the following.

For motivation, we first study the second-order derivative of a nonlinear operator. Suppose that X and Y are two normed linear spaces with an open set $\Omega \subseteq X$, and that an operator $T : \Omega \to Y$ is differentiable at $x_0 \in \Omega$. Then, T is also differentiable at every point of any open ball $B_r(x_0)$ of center x_0 and radius $r > 0$ such that $B_\rho(x_0) \subseteq \Omega$. For each fixed $x \in B_r(x_0)$, we have seen that its derivative $T'(x) := T'(x)(\cdot)$ is a bounded linear operator from X to Y, namely, an element of the family $\mathcal{L}(X, Y)$. However, if $x \in B_\rho(x_0)$ varies, $T'(x)$ is nonlinear in general, as has been seen from Example 1.23 above, so that by considering $T'(\cdot)$ as a mapping from Ω to $\mathcal{L}(X, Y)$, we can further discuss its differentiability at any point x, say x_0, in $B_r(x_0)$.

1.26. Definitions. Let X and Y be two normed linear spaces and $\Omega \subseteq X$ be an open set, and suppose that T is differentiable once at every point of the ball $B_r(x_0)$ of center x_0 and radius $r > 0$ in Ω. Then, T is said to be *twice differentiable* at x_0 if there exists a bounded linear operator $L'(x_0) : X \to Y$ such that for any $h \in X$ with $x_0 + h \in B_r(x_0)$,

$$\lim_{\|h\|_X \to 0} \frac{\|T'(x_0 + h) - T'(x_0) - L'(x_0)(h)\|_Y}{\|h\|_X} = 0 \,.$$

In this case, $L'(x_0) := L'(x_0)(\cdot)$ is called the *second derivative* of T at x_0 and is denoted by $L'(x_0) = T''(x_0)$. Moreover, if $T : \Omega \to Y$ is twice differentiable at every point $x \in \Omega$, then $L'(\cdot) = T''(\cdot)$ is a mapping from Ω to $\mathcal{L}(X, Y)$ and is called the *second-derivative mapping* of T.

We remark that as a second-derivative mapping, the operator $L'(\cdot) = T''(\cdot)$ from Ω to $\mathcal{L}(X, Y)$ is a bounded symmetric bilinear operator (see Definition 1.1). We prove this fact in the following.

1.27. Theorem. *Let* $T : X \to Y$ *be a nonlinear operator having a second-derivative at a point* $x_0 \in \Omega \subseteq X$. *Then, its second-derivative mapping* $T''(\cdot) : \Omega \to \mathcal{L}(X,Y)$ *is a bounded symmetric bilinear operator from* X^2 *to* Y.

In view of this theorem, we see that it would be more precise to write $T''(\cdot) : \Omega \to \mathcal{L}(X^2, Y)$, which we will do below.

Proof. Since the variable h in Definition 1.21 of the first derivative $T'(\cdot)$ and the variable h in Definition 1.26 of the second derivative $T''(\cdot)$ are independent, it is clear that $T''(\cdot)$ is a bounded bilinear operator from X^2 to Y. This can also be seen from the following proof of the symmetry.

Assume that $h_1, h_2 \in X$ are two nonzero elements. Suppose that $T'(x_0)$ exists in the ball $B_r(x_0)$ and that $\|h_1\|_X + \|h_2\|_X < r$. By Definition 1.21, we have

$$T'(x_0)(h_1 \to 0) = \lim_{\|h_1\|_X \to 0} \frac{T(x_0 + h_1) - T(x_0)}{\|h_1\|_X},$$

where $(h_1 \to 0)$ indicates how the limiting process is taken, and

$$T'(x_0 + h_2)(h_1 \to 0) = \lim_{\|h_1\|_X \to 0} \frac{T(x_0 + h_2 + h_1) - T(x_0 + h_2)}{\|h_1\|_X}.$$

Similarly, if $T''(x_0)$ exists, then by Definition 1.26 we have

$$T''(x_0)(h_2 \to 0, h_1 \to 0)$$
$$= \lim_{\|h_2\|_X \to 0} \frac{T'(x_0 + h_2)(h_1 \to 0) - T'(x_0)(h_1 \to 0)}{\|h_2\|_X},$$

so that

$$T''(x_0)(h_2 \to 0, h_1 \to 0)$$
$$= \lim_{\|h_2\|_X \to 0} \lim_{\|h_1\|_X \to 0}$$
$$\frac{T(x_0 + h_2 + h_1) - T(x_0 + h_2) - T(x_0 + h_1) + T(x_0)}{\|h_2\|_X \|h_1\|_X}$$
$$= \lim_{\|h_1\|_X \to 0} \lim_{\|h_2\|_X \to 0}$$
$$\frac{T(x_0 + h_1 + h_2) - T(x_0 + h_1) - T(x_0 + h_2) + T(x_0)}{\|h_1\|_X \|h_2\|_X}$$
$$= T''(x_0)(h_1 \to 0, h_2 \to 0).$$

This establishes the symmetry of the operator $T''(x_0)$ and hence completes the proof of the theorem. □

By extending the above idea, it is now possible to recursively define higher-order derivatives of a nonlinear operator $T : \Omega \to Y$.

1.28. Definitions. Let X and Y be two normed linear spaces and $\Omega \subseteq X$ be an open set. Suppose that T is continuously differentiable $(n-1)$ times at every point of the ball $B_r(x_0)$ of center x_0 and radius $r > 0$, where $B_\rho(x_0) \subseteq \Omega$. Then, T is said to be *differentiable n times* at x_0 if there exists a bounded linear operator $L^{(n-1)}(x_0) : X \to Y$ such that for any $h \in X$ with $x_0 + h \in B_r(x_0)$,

$$\lim_{\|h\|_X \to 0} \frac{\|T^{(n-1)}(x_0 + h) - T^{(n-1)}(x_0) - L^{(n-1)}(x_0)(h)\|_Y}{\|h\|_X} = 0.$$

In this case, $L^{(n-1)}(x_0)$ is called the nth *derivative* of T at x_0 and is denoted by $L^{(n-1)}(x_0) = T^{(n)}(x_0)$. Moreover, if $T : \Omega \to Y$ is differentiable n times at every point $x \in \Omega$, then $L^{(n-1)}(\cdot) = T^{(n)}(\cdot)$ is a mapping from Ω to $\mathcal{L}(X, Y)$ and is called the *nth-derivative mapping* of T.

Similar to Theorem 1.27, we have the following result.

1.29. Theorem. *The nth-derivative mapping (if it exists) $T^{(n)}(\cdot) : \Omega \to \mathcal{L}(X, Y)$ is a bounded symmetric n-linear operator from X^n to Y.*

Again, in view of this theorem, it would be more precise to write $T^{(n)}(\cdot) : \Omega \to \mathcal{L}(X^n, Y)$, which we will do below.

Now we are in a position to introduce the notion of operator-valued Taylor series. First, let us consider the nth-order polynomic operator $P_n : X \to Y$ defined by n bounded k-power operators (see Theorem 1.10) as follows:

1.30. Definition. Consider an operator $P_n : X \to Y$. If there exist n bounded k-power operators $a_k[\cdot] : X \to Y$, $k = 1, \ldots, n$, and an element $a_0 \in Y$ such that

$$P_n(x) = a_n[x^n] + \cdots + a_2[x^2] + a_1[x] + a_0$$

for all $x \in X$, then $P_n(\cdot)$ is called an *nth-order polynomic operator* defined on X.

Here, as remarked after Definition 1.3, the domain of the k-power operator $a_k[\cdot]$ is actually the diagonal of $X^k, k = 1, \ldots, n$. Since each coordinate axis of the diagonal of X^k is identical to X, $a_k[\cdot]$ can be considered to be defined on X as well.

By Theorem 1.6, there is a unique symmetric bounded k-linear operator $a_k(\cdot) : X^k \to Y$ corresponding to $a_k[\cdot]$ for each $k, k = 1, 2, \ldots, n$. It follows from Definitions 1.21, 1.26, and 1.28 that P_n is infinitely many times differentiable on X with

$$
\left\{
\begin{array}{l}
P_n'(x) = na_n[x^{n-1}] + \cdots + 2a_2[x] + a_1\,, \\[4pt]
\quad \vdots \\[4pt]
P_n^{(n)}(x) = n!a_n\,, \\[4pt]
P_n^{(n+1)}(x) = P_n^{(n+2)}(x) = \cdots = 0\,,
\end{array}
\right.
\tag{1.7}
$$

where $x \in X$. Here, the notation must be clarified. Let us consider, for example, the formula $P_n'(x)$ in (1.7). Since $a_1[\cdot] = a_1(\cdot) : X \to Y$ is a bounded linear operator by definition, it follows from Theorem 1.22, part (2), that $a_1'[x_0] = a_1$ for any $x_0 \in X$, so that the symbol a_1 in the expression of $P_n'(x)$ represents a bounded linear operator (recall that $P_n'(x)$ is a bounded linear operator for each fixed $x \in X$). This is different from the symbol a_0 in $P_n(x)$, which is an element in Y. On the other hand, it can be similarly verified that $a_2[x]$ is also a bounded linear operator from X to Y for each fixed x. The same interpretation applies to each term of P_n'. Consequently, $P_n'(x)(\cdot)$ is a bounded linear operator from X to Y for each fixed $x \in X$. It is easy to see, in the same manner, that $P_n^{(n)}(x) = n!a_n$ is a bounded linear operator from X to Y for each fixed $x \in X$. With these notations, we have the following:

1.31. Definition. Let $x_0 \in X$ be fixed. An operator $P_n : X \to Y$ is said to have a finite Taylor series expansion at x_0 if its derivatives $P_n^{(k)}(x_0)$ exist for $k = 1, \ldots, n$, such that

$$
\begin{aligned}
&P_n(x + x_0) \\
&= P_n[x_0] + P_n'(x_0)[x] + \frac{1}{2!}P''(x_0)[x^2] + \cdots + \frac{1}{n!}P_n^{(n)}(x_0)[x^n]\,.
\end{aligned}
$$

Comparing it with Definition 1.30, we see that if a nonlinear operator $T : X \to Y$ has a finite Taylor series expansion at $x_0 \in X$, then it is a polynomic operator of the same order. More precisely, in this case we have

$$a_k = \frac{1}{k!}T^{(k)}(x_0), \quad k = 0, 1, \ldots, n,$$

where $a_k[\cdot] : X \to Y$ is a bounded k-power operator, $k = 1, \cdots, n$, with $a_0 = T(x_0) \in Y$.

Recall Theorem 1.25. In particular, let the nonlinear operator T stated therein be a bounded n-power operator from X to Y. Then, it is easily seen that T is differentiable if and only if T is continuous. Namely, we have the following theorem.

1.32. Theorem. *Let* $T : X \to Y$ *be a bounded n-power operator. Then,* T *is differentiable if and only if T is continuous.*

Proof. The necessity is obvious. To show the sufficiency, observe that the bounded linear operator $L(x_0) = nT[x_0^{n-1}]$ always exists for any given $x_0 \in X$, so that the sufficient conditions of Theorem 1.25 are satisfied. $\qquad\square$

Consequently, we have the following important result:

1.33. Theorem. *Let* Ω *be an open subset of X and suppose that the nth-derivative* $T^{(n)}(x)$ *of a nonlinear operator* $T : \Omega \to Y$ *is continuous on* Ω. *Then, for any $x, h \in X$ such that $x + h \in \Omega$, we have*

$$T(x + h) = T(x) + T'(x)[h] + \frac{1}{2!}T''(x)[h^2]$$

$$+ \cdots + \frac{1}{(n-1)!}T^{(n-1)}(x)[h^{n-1}] + o(\|h\|_X^n).$$

This is usually called the *Taylor expansion* for differentiable nonlinear operators.

Proof. For $n = 1$, this is just the definition of the derivative (see Definition 1.21). The proof can then be completed by mathematical induction with a repeated use of the definition. $\qquad\square$

§ 7. Infinite Power Series

In this section, we are concerned with infinite power series expansions of certain nonlinear operators.

Again, let X be a normed linear space but let Y be a Banach space. Consider the formal Fréchet power series, or simply power series, of an operator $T : X \to Y$ of the form

$$T(x) = \sum_{n=0}^{\infty} a_n[x^n], \quad x \in X, \tag{1.8}$$

where $a_n : X \to Y$ is a bounded n-power operator and the equality is understood in the same way as that discussed in Section 6 for finite Taylor series. We first have the following definitions.

1.34. Definitions. The formal power series (1.8) is said to be *absolutely convergent* at a point $x_0 \in X$ if

$$\sum_{n=0}^{\infty} \|a_n[x_0^n]\|_Y < \infty.$$

If, furthermore, series (1.8) is absolutely convergent at every point of a subset B of X, then it is said to be *absolutely convergent* on B.

In what follows, we will be mainly concerned with the above-defined convergence, unless otherwise stated, and hence will simply call an absolutely convergent power series *convergent*. Moreover, if the power series (1.8) converges and is unique, then we will say that the operator T has an *infinite power series representation* (1.8). The uniqueness of the power series representation of a nonlinear operator $T : X \to Y$ will be discussed in the next section. Here, we only consider the existence (or convergence) issue.

1.35. Theorem. *Suppose that* $T : X \to Y$ *has a formal power series* (1.8). *For any* $x_0 \in X$ *with* $\|x_0\|_X = 1$, *define*

$$\rho(x_0) = \overline{\lim_{n\to\infty}} \, \|a_n[x_0^n]\|_Y^{1/n}.$$

Then, the power series (1.8) *converges inside an open star-shaped set of center 0 and radius* $r = 1/\rho(x_0)$ *and diverges outside the corresponding closed star-shaped set of center 0 and radius* $r = 1/\rho(x_0)$.

Proof. Let $x = \lambda x_0$ where $\|x_0\|_X = 1$. If $\lambda < 1/\rho(x_0)$, then we have

$$\overline{\lim_{n \to \infty}} \|a_n[x^n]\|_Y^{1/n} = \overline{\lim_{n \to \infty}} \|a_n[\lambda^n x_0^n]\|_Y^{1/n} = |\lambda|\rho(x_0) < 1.$$

Hence, there exist an integer N and an $r < 1$ such that for $n > N$,

$$\|a_n[x^n]\|_Y^{1/n} \leq r < 1 \quad \text{or} \quad \|a_n[x^n]\|_Y \leq r^n < 1.$$

Since Y is a Banach space and is hence complete in norm, by the M-test in calculus, the (power) series $\sum_{n=0}^{\infty} \|a_n[x^n]\|_Y$ converges. If $\lambda > 1/\rho(x_0)$, then similarly we have

$$\overline{\lim_{n \to \infty}} \|a_n[x^n]\|_Y^{1/n} > 1,$$

so that by the definition of the sup-limit there are infinitely many integers n such that $\|a_n[x^n]\|_Y > 1$. Consequently, the power series diverges, completing the proof of the theorem. □

We remark that unlike the ordinary power series, the convergent region of the power series (1.8) is in general only a star-shaped set, as stated and proved in the above theorem, and is not necessarily a ball in the normed linear space X. This can also be seen from the following example.

1.36. Example. Let $X = \ell_\infty$ and consider the power series

$$\sum_{n=0}^{\infty} a_n[x^n] = 1 + \sum_{n=1}^{\infty} x_1 x_2 \cdots x_n,$$

where the sequence $(x_1, x_2, \dots) \in X$. Obviously, the n-power operator $a_n[\cdot] : X \to R$, defined by $a_n[x^n] = x_1 x_2 \cdots x_n$ (see Example 1.4) has an operator norm 1. This power series converges at the points

$$x^{(1)} = (2^{-1}, 2^{-1}, 2^{-1}, \cdots) \quad \text{and} \quad x^{(2)} = (0, 2, 2, \cdots),$$

but not at the point $\frac{1}{2}(x^{(1)} + x^{(2)})$. Hence, the region of convergence for this power series is not even a convex set.

However, in the following case, we may as usual have an open ball for the region of convergence for the infinite power series (1.8).

1.37. Corollary. *Let $\rho(x_0)$ be defined as in Theorem 1.35, and let*

$$\rho_0 = \sup_{\|x_0\|_X = 1} \rho(x_0).$$

Then, the power series (1.8) *converges inside the open ball of radius $r < 1/\rho_0$. However, for any $r > 1/\rho_0$, the power series* (1.8) *diverges at some point x with $\|x\|_X = r$.*

Proof. The power series converges at $x = 0$. For any $x \in X$ with $0 < \|x\|_X < 1/\rho_0$, let $x_0 = x/\|x\|_X$. Then $x = \|x\|_X x_0$, where $\|x\|_X < 1/\rho_0 \le 1/\rho(x_0)$, so that by Theorem 1.35 the power series converges inside the open ball of radius $r < 1/\rho_0$. If $r > 1/\rho_0$, then there exists an x_0 with $\|x_0\|_X = 1$ such that $r > 1/\rho(x_0)$. By Theorem 1.35, the power series diverges at the point $x = rx_0$. $\square$

Note that in the above discussion the convergence is pointwise and absolute. The following result gives a criterion for *uniform convergence*.

1.38. Theorem. *Suppose that an operator $T : X \to Y$ has a formal power series* (1.8). *Let*

$$\rho_T = \varlimsup_{n \to \infty} \|a_n\|^{1/n},$$

where $\|a_n\|$ is the operator norm of the n-power operator $a_n[\cdot] : X \to Y$ defined as in 1.9. Then, the power series (1.8) *converges absolutely and uniformly on any closed ball of center 0 and radius $r < 1/\rho_T$. However, for any $r > 1/\rho_T$, the series* (1.8) *does not converge uniformly on the set*

$$S = \{x \in X : \|x\|_X = r\}.$$

Proof. Suppose that $r < 1/\rho_T$. Then there exists an $\epsilon > 0$ such that $r = (1 - \epsilon)/\rho_T$, and there is an integer N such that for $n > N$ we have $\|a_n\|^{1/n} < (1+\epsilon)\rho_T$. Consequently for all x with $\|x\|_X \le r$,

$$\|a_n[x^n]\|_Y \le \|a_n\| \, \|x\|_X^n < (1 + \epsilon)^n \rho_T^n \frac{(1 - \epsilon)^n}{\rho_T^n} = (1 - \epsilon^2)^n.$$

This implies that the series $\sum_{n=0}^{\infty} \|a_n[x^n]\|_Y$ converges absolutely and uniformly on the closed ball of center 0 and radius r, and so does series (1.8)

Conversely, suppose that $r > 1/\rho_T$. Then, there exists an $\epsilon > 0$ such that $r > 1/((1-\epsilon)\rho_T)$, and for any n there is an x_n with $\|x_n\|_X = 1$ such that $\|a_n[x_n^n]\|_Y^{1/n} \geq (1-\epsilon)\rho_T$. Consequently, for this x_n we have

$$\|a_n[(rx_n)^n]\|_Y = r^n \|a_n[x_n^n]\|_Y \geq r^n (1-\epsilon)^n \rho_T^n > 1,$$

so that for any n,

$$\sup_{\|x\|_X = r} \|a_n[x^n]\|_Y \geq 1.$$

This implies that the limit does not tend to zero uniformly on the set $S = \{x \in X : \|x\|_X = r\}$. Hence, series (1.8) does not converge uniformly on the set S. $\square$

We close this section with some further discussions.

1.39. Definition. A sequence of non-negative real numbers

$$\{\, \alpha_0, \alpha_1, \alpha_2, \dots \,\}$$

is called a *majorizing sequence* for the infinite power series (1.8) if

$$\|a_n\| \leq \alpha_n \quad \text{for all } n, \quad n = 0, 1, 2, \dots,$$

where $\|a_n\|$ is the operator norm of the n-power operator $a_n[\cdot]$: $X \to Y$, defined as in Theorem 1.9.

By the M-test in calculus, the following result is immediate.

1.40. Theorem. *Suppose that the power series (1.8) has a majorizing sequence $\{\alpha_0, \alpha_1, \alpha_2, \dots\}$ with $\overline{\lim}_{n\to\infty} \alpha_n^{1/n} = r < \infty$. Then the power series (1.8) converges absolutely and uniformly on any closed ball of center 0 and radius $\rho < 1/r$.*

Note that in case $r = 0$ in the above theorem, the power series (1.8) converges absolutely and uniformly on the entire normed linear space X.

We finally remark that, as mentioned above, the uniqueness issue for the power series (1.8) will be studied in the next section.

§ 8. Nonlinear Analytic Mappings

In this section, we again let X be a normed linear space and Y a Banach space.

1.41. Definitions. Let $T : X \to Y$ be a nonlinear operator defined on an open subset $\Omega \subseteq X$. If T has a unique infinite power series representation

$$T(x) = \sum_{n=0}^{\infty} a_n[x^n], \quad x \in \Omega, \tag{1.9}$$

which converges absolutely on Ω and uniformly on every closed bounded subset of Ω, then T is called an *analytic mapping* (or *analytic operator*) from Ω to Y. In this case, T is also said to be *analytic on Ω.*

As has been seen from Theorem 1.40, if the operator norm $\|a_n\| \leq \alpha_n$ for all n and if $\overline{\lim}_{n \to \infty} \alpha_n^{1/n} = r < \infty$, then the infinite power series representation (1.9) converges absolutely on a ball of center 0 and radius at least equal to $1/r$, and converges absolutely and uniformly on every closed bounded subset of the ball.

We will study the continuity and differentiability of an analytic mapping $T : \Omega \to Y$. The continuity is obvious. More precisely, we have the following:

1.42. Theorem. *Let X be a normed linear space and Y be a Banach space. If $T : \Omega \to Y$ is analytic on an open subset $\Omega \subseteq X$, then T is continuous on Ω.*

Proof. We show that for any point $x_0 \in \Omega$, T is continuous at x_0. Let B_δ be a closed ball centered at x_0 with radius $\delta > 0$ such that $B_\delta \subset \Omega$. Then, by definition T converges absolutely and uniformly on B_δ. Hence, for any $\epsilon > 0$, there exists an integer N such that for all $x \in B_\delta$,

$$\left\| T(x) - \sum_{n=0}^{N} a_n[x^n] \right\|_Y < \frac{\epsilon}{3}.$$

Since the finite sum $\sum_{n=0}^{N} a_n[x^n]$ is continuous, for all $x \in B_\delta$ we have

$$\left\| \sum_{n=0}^{N} a_n[x^n] - \sum_{n=0}^{N} a_n[x_0^n] \right\|_Y < \frac{\epsilon}{3},$$

so that for all $x \in B_\delta$,

$$\|T(x) - T(x_0)\|_Y$$

$$\leq \left\| T(x) - \sum_{n=0}^{N} a_n[x^n] \right\|_Y + \left\| \sum_{n=0}^{N} a_n[x^n] - \sum_{n=0}^{N} a_n[x_0^n] \right\|_Y$$

$$+ \left\| \sum_{n=0}^{N} a_n[x_0^n] - T(x_0) \right\|_Y < \epsilon.$$

This implies that the analytic mapping T is continuous at x_0, completing the proof of the theorem.   □

The general result on the differentiability of $T : \Omega \to Y$ is relatively complicated. The following theorem, however, can be established by imitating the same argument as that for scalar power series in calculus.

1.43. Theorem. *Let X be a normed linear space and Y be a Banach space. Let $T : B_r \to Y$ be analytic on the ball B_r of center 0 with radius $r > 0$ in X. Then, T is differentiable on B_r, with*

$$T'(x) = \sum_{n=1}^{\infty} n a_n[x^{n-1}], \quad x \in B_r, \tag{1.10}$$

where $T'(x)$ is also analytic on a ball of center 0 with radius at least equal to r.

Proof. Let

$$\rho = \varlimsup_{n \to \infty} \|n a_n\|^{1/n},$$

where $\| \cdot \|$ denotes the operator norm as usual. We will show that $(1/\rho) \geq r$, where r is the radius of the ball B_r on which the mapping

T is analytic, so that by Theorem 1.38 the power series (1.10) converges absolutely and uniformly on every closed ball of center 0 with a radius $< r$. To this end, the proof of the theorem can be completed by imitating the same argument as that for standard scalar power series in calculus.

To show that $(1/\rho) \geq r$, we first observe that $a_n[x^{n-1}h]$ is a linear operator in $h \in X$. For any fixed integer $n, n = 1, 2, \ldots$, let $x, h \in X$ be such that $\|x\|_X = 1$ and $\|h\|_X = 1/n$. Then, by the Hahn-Banach theorem, there exists a functional $y^* \in Y^*$, the dual space of Y, with $\|y^*\| = 1$ such that

$$y^*\big(na_n[x^{n-1}h]\big) = \|na_n[x^{n-1}h]\|_Y \,.$$

Let

$$f(\lambda) = y^*\big(a_n[(x + \lambda h)^n]\big) \,.$$

Then, for $-1 \leq \lambda \leq 1$, we have

$$|f(\lambda)| \leq \|y^*\| \, \|a_n\| \, (\|x\|_X + |\lambda|\|h\|_X)^n \leq \|a_n\|\left(1 + \frac{|\lambda|}{n}\right)^n \leq 3\|a_n\| \,,$$

so that

$$\sup_{-1\leq\lambda\leq1} |f(\lambda)| \leq 3\|a_n\| < \infty \,.$$

Let $g(\lambda) = a_n[(x + \lambda h)^n]$. Since $g(\lambda)$ is a polynomial of degree n in λ, by the Markov inequality (see Reference [7]), we have

$$\sup_{-1\leq\lambda\leq1}\left|\frac{d}{d\lambda}g(\lambda)\right| \leq n^2 \sup_{-1\leq\lambda\leq1} |g(\lambda)| \leq 3n^2\|a_n\| \,.$$

Consequently, since $g'(0) = na_n[x^{n-1}h]$, we have

$$\|na_n[x^{n-1}h]\|_Y \leq 3n^2\|a_n\| \,.$$

It then follows that the operator norm (for the linear operator $a_n[x^{n-1}h]$ in the variable h)

$$\|na_n[x^{n-1}]\|_h := \sup_{\|h\|_X=1} \|na_n[x^{n-1}h]\|_Y$$

$$= \sup_{\|h\|_X=\frac{1}{n}} \frac{\|na_n[x^{n-1}h]\|_Y}{\|h\|}$$

$$\leq 3n^3\|a_n\| \,,$$

so that the operator norm (for the $(n-1)$-linear operator $a_n[x^{n-1}h]$ in the variable x^{n-1})

$$\|na_n[h]\|_{x^{n-1}} := \sup_{\|x\|_X=1} \|na_n[x^{n-1}h]\|_Y$$

$$\leq \|na_n[x^{n-1}]\|_h \|h\|_X$$

$$\leq 3n^2 \|a_n\|.$$

Therefore, we have

$$\rho = \varlimsup_{n\to\infty} \|na_n\|^{1/n}$$

$$= \varlimsup_{n\to\infty} \|na_n[h]\|_{x^{n-1}}^{1/n}$$

$$\leq \varlimsup_{n\to\infty} \left(3n^2 \|a_n\|\right)^{1/n}$$

$$= \varlimsup_{n\to\infty} \|a_n\|^{1/n},$$

or $(1/\rho) \geq r$. This establishes the radius of convergence for the power series (1.10) and hence completes the proof of the theorem. $\square$

Finally, we have the following important result, which is an immediate consequence of the above theorem and is a generalization of the well-known result for analytic functions in complex analysis.

1.44. Corollary. *Let X be a normed linear space and Y be a Banach space. Then, an analytic mapping $T : B \to Y$ defined on an open ball $B \subseteq X$ is infinitely many times differentiable on B. Moreover, the bounded n-power operators*

$$a_n[\cdot] = \frac{1}{n!} T^{(n)}(0)(\cdot), \quad n = 0, 1, 2, \ldots,$$

are uniquely determined by the given mapping T.

§ 9. Nonlinear Volterra Mappings

In this section, we study a special, yet very important, type of non-linear analytic mappings: the nonlinear Volterra mappings. This particular type of nonlinear analytic mappings has many important applications in nonlinear systems control engineering and signal processing.

Let $X = L_1(0,T)$, and let $Y = C(0,T)$ be equipped with the sup-norm, where $T \leq \infty$. For each fixed $n \geq 1$, let $K_n = K_n(t; t_1, \ldots, t_n)$ be a complex-valued function with the property that $K_n(\cdot; t_1, \ldots, t_n) \in L_1(0,T)$ for all $t_1, \ldots, t_n \in [0,T)$ and $K_n(t; \cdot, \ldots, \cdot) \in L_1([0,T) \times \cdots \times [0,T))$ for all $t \in [0,T)$. Then, the operator $V_n : X \to Y$ defined by

$$V_n(\cdot)(t) := \int_0^t \cdots \int_0^{t_2} K_n(t; t_1, \ldots, t_n)(\cdot)(t_1) \cdots (\cdot)(t_n) dt_1 \cdots dt_n$$

$$(1.11)$$

is a bounded and symmetric n-linear operator, which maps an element $x(t) \in X$ to an element in Y expressed as

$$y(t) = V_n(x)(t)$$
$$= \int_0^t \cdots \int_0^{t_2} K_n(t; t_1, \ldots, t_n) x(t_1) \cdots x(t_n) dt_1 \cdots dt_n \,.$$

Introducing a notation for the so-called *plus function*,

$$(t - t_0)_+^k = \begin{cases} (t - t_0)^k & \text{if } t > t_0 \\ 0 & \text{if } t \leq t_0, \end{cases}$$

we may rewrite (1.11) as follows:

$$V_n(\cdot)(t) = \int_0^T \cdots \int_0^T \tilde{K}_n(t; t_1, \ldots, t_n)(\cdot)(t_1) \cdots (\cdot)(t_n) dt_1 \cdots dt_n \,,$$

$$(1.12)$$

where

$$\tilde{K}_n(t; t_1, \ldots, t_n)$$
$$= K_n(t; t_1, \ldots, t_n)(t_2 - t_1)_+^0 \cdots (t_n - t_{n-1})_+^0 (t - t_n)_+^0 \,.$$

Because of this equivalence, we will always use the notation K_n to mean $\tilde{K}_n$ below, unless otherwise indicated.

Now, let Ω be the open ball of center 0 and radius $\delta > 0$ in $X = L_1(0,T)$ and suppose that a function $V_0(t) \in Y$ and all operators $V_n(\cdot)$ have been defined on Ω as above, $n = 1, 2, \ldots$. Assume, furthermore, that there is a majorizing sequence $\{\alpha_0, \alpha_1, \alpha_2, \ldots\}$ with $\overline{\lim}_{n\to\infty} \alpha_n^{1/n} < \infty$, such that

$$\|V_n\| \leq \alpha_n \quad \text{for all } n, \quad n = 0, 1, 2, \ldots,$$

where $\|V_n\|$ is the operator norm of $V_n : \Omega \to Y$. Let

$$r = \varlimsup_{n\to\infty} \alpha_n^{1/n}.$$

Then, the infinite power series

$$V(\cdot) := \sum_{n=0}^{\infty} V_n(\cdot)(t) = V_0(t) + \sum_{n=1}^{\infty} \int_0^T \cdots \int_0^T$$
$$K_n(t; t_1, \ldots, t_n)(\cdot)(t_1) \cdots (\cdot)(t_n) dt_1 \cdots dt_n \quad (1.13)$$

converges absolutely and uniformly on a ball of radius at least equal to ρ where $\rho = min\{\delta, 1/r\}$ (see Theorem 1.40).

1.45. Definitions. An (absolutely and uniformly) convergent infinite series of the form (1.13) is called a (nonlinear) *Volterra series*, and the operator $V(\cdot)$ defined by such a Volterra series is called a (nonlinear) *Volterra mapping*. Moreover, $\{K_n(t; t_1, \ldots, t_n)\}$ are called the *Volterra kernels* of the mapping $V(\cdot)$.

In particular, when $V_n \equiv 0$ for all $n \neq 1$, the Volterra mapping $V(\cdot)$ is *linear*.

1.46. Definition. If the Volterra mapping $V(\cdot)$ has the following expression:

$$V(\cdot) = V_0(t)$$
$$+ \sum_{n=1}^{\infty} \frac{w^n}{n!} \int_0^T \cdots \int_0^T K_n(t; t_1, \ldots, t_n)(\cdot)(t_1) \cdots (\cdot)(t_n) dt_1 \cdots dt_n$$
$$:= \sum_{n=0}^{\infty} \frac{w^n}{n!} \int_{[0,T]^n} K_n(t; t_1, \ldots, t_n)(\cdot)(t_1) \cdots (\cdot)(t_n) dt_1 \cdots dt_n,$$

where w is a real number (need not be positive in general) and $K_0(t) := V_0(t)$, then $V(\cdot)$ is called a *w-weighted Volterra mapping*.

The reason for studying weighted Volterra mappings is clear. For example, if $T < \infty$ and all the Volterra kernels

$$K_n(t; t_1, \ldots, t_n), \quad n = 0, 1, 2, \ldots,$$

are bounded in the sense that

$$|K_n(t; t_1, \ldots, t_n)| \le M^n, \quad t \in [0, T], \quad n = 0, 1, 2, \ldots,$$

for some constant $M < \infty$, then we have

$$|V(x)(t)| \le \sum_{n=0}^{\infty} \frac{(wM\delta T)^n}{n!} = e^{wM\delta T} < \infty$$

for all $x(t) \in \Omega$ and all $t \in [0, T]$, where δ is the radius of the ball Ω on which the Volterra mapping $V(\cdot)$ is defined. Consequently, the infinite Volterra series can be approximated by its finite truncations such that the approximation error is within an arbitrarily small tolerance. In doing so, w can be used to control the convergence rate in the approximation. This is very convenient in engineering applications.

Exercises

1.1. Give some examples of symmetric and nonsymmetric n-linear operators that are different (in form) from the Volterra operator and those shown in Example 1.2.

1.2. Give an example to show that two different n-linear operators can determine the same n-power operator, while an n-power operator can be used to define two different nonsymmetric n-linear operators.

1.3. Let Δ be a "difference" operator defined inductively on the n-power operator $f_n[\cdot]$ as follows:

$$\begin{cases} \Delta^1_{x_1} f[x] = f[x + x_1] - f[x], \\ \Delta^n_{x_1,\ldots,x_n} f[x] = \Delta^1_{x_n}\left(\Delta^{n-1}_{x_1,\ldots,x_{n-1}} f[x]\right), \quad n = 2, 3, \ldots. \end{cases}$$

Show that

(1)

$$\Delta^n_{x_1,\ldots,x_n} f_n[x^n] = (-1)^n f_n[x^n]$$
$$+ \sum_{k=1}^{n} (-1)^{n-k} \sum_{1 \leq j_1,\ldots,j_k \leq n} f_n[(x + x_{j_1} + \cdots + x_{j_k})^n].$$

This completes the proof of Theorem 1.7.

(2) If $x = -\frac{1}{2}\sum_{i=1}^{n} x_i$, then

$$\Delta^n_{x,\ldots,x} f_n[x^n] = \frac{1}{2^n n!}\left\{ f_n\left[\left(\sum_{i=1}^{n} x_i\right)^n\right]\right.$$
$$\left. + \sum_{k=1}^{n} (-1)^{n-k} \sum_{e_j = \pm 1} f_n\left[\left(\sum_{i=1}^{n} e_i x_i\right)^n\right]\right\},$$

where $\sum_{e_j = \pm 1}$ has a total of $\binom{n}{k}$ terms. This completes the proof of Theorem 1.10.

1.4. Give simple yet nontrivial examples to show that a continuous nonlinear operator can be unbounded.

1.5. Show that the two-variable nonlinear function $f : R^2 \to R$ defined by

$$f(x_1, x_2) = \begin{cases} \dfrac{2x_1 x_2}{x_1^2 + x_2^2} & \text{if } x_1^2 + x_2^2 \neq 0 \\ \\ 0 & \text{if } x_1^2 + x_2^2 = 0 \end{cases}$$

is continuous in each $x_i, i = 1, 2$, but is not continuous in the 2-tuple $\mathbf{x} = (x_1, x_2)$.

1.6. Verify the results stated in Theorem 1.22.

1.7. Give some examples of analytic mappings that are not polynomic or Volterra operators.

References

[1] A. Alexiewicz and W. Orlicz, "Analytic operations in real Banach spaces," *Studia Mathematica*, **14** (1953), 57−78.

[2] M.S. Berger, *Nonlinearity and Functional Analysis*, Academic Press, New York, 1977.

[3] J. Bochnak and J. Siciak, "Polynomials and multilinear mappings in topological vector spaces," *Studia Mathematica*, **36** (1971), 59−76.

[4] N. Bourbaki, *Elèments de Mathèmatique: Espaces vectoriels topologiques*, Fasc. 18, Livre 6, Hermann, Paris, 1967.

[5] V.J. Bruno, "The continuity of increasing polynomial operators on ordered topological vector spaces," *Nonlinear Analysis: Theory, Methods & Applications*, **4** (1980), 815−820.

[6] W.Y. Chen, *Nonlinear Functional Analysis* (in Chinese), Ganshu Pub., China, 1982.

[7] E. Cheney, *Introduction to Approximation Theory*. McGraw-Hill, New York, 1966.

[8] R.J.P. de Figueiredo and T.A.W. Dwyer, "A best approximation framework and implementation for simulation of large-scale nonlinear systems," *IEEE Trans. Circ. Sys.*, **27** (1980), 1005−1014.

[9] R.J.P. de Figueiredo and G. Chen, "Optimal interpolation on a generalized Fock space of analytic functions," in *Approximation Theory* (C.K. Chui *et al.*, eds.), pp. 247−250, Academic Press, New York, 1989.

[10] A. Halme, "Polynomial operators for nonlinear systems analysis," *Acta Polytechnica Scandinavica*, **24** (1972), 1−64.

[11] E. Hille and R.S. Phillips, *Functional Analysis and Semigroups*, AMS Colloq. Pub., New York, 1957.

[12] D.H. Hyers, "Polynomial operators," in *Topics in Mathematical Analysis* (Th.M. Rassias ed.), pp. 410−444, World Sci. Pub., Singapore, 1989.

[13] M.A. Krasnoselskij and G.M. Vainikko, *et al.*, *Approximate Solutions of Operator Equations*, Wolters-Noordhoff, Groningen, 1972.

[14] G. Pólya and G. Szegö, *Problems and Theorems in Analysis*, Springer-Verlag, New York, Vol. 1, 1972, and Vol. 2, 1976.

[15] G. Prodi and A. Ambrosetti, *Analisi Non Lineare*, Quaderno Scuola Normale Superiore, Pisa, 1973.

[16] L.B. Rall, *Computational Solutions of Nonlinear Operator Equations*, Wiley, New York, 1969.

[17] W.J. Rugh, *Nonlinear System Theory: The Volterra/Wiener Approach*, Johns Hopkins Univ. Press, Baltimore, 1981.

[18] V. Volterra, *Theory of Functionals and of Integrals and Integodifferential Equations*, Dover Pub., New York, 1959.

[19] L.V. Zyla, "Nonlinear system approximation and identification," Ph.D. thesis, Department of Electrical and Computer Engineering, Rice University, Houston, Texas, 1977.

Chapter 2

Nonlinear Lipschitz Operators

Nonlinear Lipschitz operators from a normed linear space to another normed linear space (of complex-valued functions defined on the time domain) are the fundamental entities on which the results in the following chapters are based. For this reason, we devote this chapter to them. As is suggested by its name, the concept of a nonlinear Lipschitz operator constitutes a natural generalization of the classical notion of a Lipschitz function. The basic topics on the nonlinear Lipschitz operators to be studied in the present chapter include a preamble on general nonlinear operators, an elementary development of the basic theory of nonlinear Lipschitz operators, the important notion of contraction mappings and the fixed point theorem, generalized nonlinear Lipschitz operators, and differentiable nonlinear mappings. Since a nonlinear analytic mapping studied in the previous chapter, including nonlinear Volterra mappings, is a Lipschitz operator under certain conditions, the general theory developed in this chapter is very useful in nonlinear systems control theory and engineering. Some applications of the generalized nonlinear Lipschitz operators to optimal design of nonlinear feedback systems will be discussed in the next few chapters.

§ 1. Preliminaries

In this section, we introduce some necessary notations and definitions that will be used throughout the chapter.

Let X and Y be linear function spaces over the field C of complex numbers. Let $A : X \to Y$ be an operator mapping from X
to Y and denote by $\mathcal{D}(A)$ and $\mathcal{R}(A)$, respectively, the *domain* and
range of A. Assume that $\mathcal{D}(A) \subseteq X$ and $\mathcal{R}(A) \subseteq Y$.

As usual, an operator $A : \mathcal{D}(A) \to Y$ is *linear* if $\mathcal{D}(A)$ is a
linear subspace of X and

$$A : \ ax_1 + bx_2 \ \to \ aA(x_1) + bA(x_2)$$

for all $x_1, x_2 \in \mathcal{D}(A)$ and all $a, b \in C$; otherwise it is said to be
nonlinear. Since linearity is a special case of nonlinearity, in what
follows "nonlinear" will always mean "not necessarily linear" unless
otherwise indicated.

For any subsets $\Omega \subseteq \mathcal{D}(A)$ and $\Sigma \subseteq Y$, we use the notation

$$A(\Omega) = \{ \ A(x) \in Y : \quad x \in \Omega \ \} \tag{2.1}$$

and

$$A^{-1}(\Sigma) = \{ \ x \in \mathcal{D}(A) : \quad A(x) \in \Sigma \ \}. \tag{2.2}$$

Here, $A^{-1}(\Sigma)$ denotes the pre-image of the set $\Sigma \subseteq Y$, and $A^{-1}(\cdot)$
does not necessarily mean an inverse operator of A, unless cer-
tain conditions are satisfied as discussed later in this section. Let
$\mathcal{N}(X,Y)$ be the family of all nonlinear operators A mapping from
its domain $\mathcal{D}(A) \subseteq X$ into Y. Recall from (1.3) that $\mathcal{L}(X,Y)$ is
used to denote the family of bounded linear operators from X to
Y. Obviously, $\mathcal{L}(X,Y) \subset \mathcal{N}(X,Y)$. In the case that $X = Y$, we
use the notation $\mathcal{L}(X)$ and $\mathcal{N}(X)$, respectively, instead of $\mathcal{L}(X,X)$
and $\mathcal{N}(X,X)$ for simplicity.

We now introduce algebraic operations into $\mathcal{N}(X,Y)$. If $A \in
\mathcal{N}(X,Y)$ and $a \in C$, then aA is the member of $\mathcal{N}(X,Y)$ defined by

$$\mathcal{D}(aA) = \mathcal{D}(A) \quad \text{and} \quad (aA)(x) = aA(x) \tag{2.3}$$

for all $x \in \mathcal{D}(aA)$. Let $A, B \in \mathcal{N}(X, Y)$. Then, if $\mathcal{D}(A) \cap \mathcal{D}(B) \neq \emptyset$, $A + B$ is the member of $\mathcal{N}(X, Y)$ defined by

$$\mathcal{D}(A + B) = \mathcal{D}(A) \cap \mathcal{D}(B) \text{ and } (A + B)(x) = A(x) + B(x) \quad (2.4)$$

for all $x \in \mathcal{D}(A+B)$. Moreover, if Z is a linear space over the field $\mathcal{C}$ and $C \in \mathcal{N}(Y, Z)$ is such that $\mathcal{R}(A) \cap \mathcal{D}(C) \neq \emptyset$, then we introduce an operation $C \circ A$ (or simply $C(A(\cdot))$ or CA) into $\mathcal{N}(X, Z)$ by defining $C \circ A$ to be the member of $\mathcal{N}(X, Z)$ satisfying

$$\mathcal{D}(C \circ A) = A^{-1}(\mathcal{R}(A) \cap \mathcal{D}(C)) \text{ and } (C \circ A)(x) = C(A(x)) \quad (2.5)$$

for all $x \in \mathcal{D}(C \circ A)$.

For any subset $D \subseteq X$, let $\mathcal{F}(D, Y)$ be the family of nonlinear operators A such that $\mathcal{D}(A) = D$ and $\mathcal{R}(A) \subseteq Y$. Clearly, $\mathcal{F}(D, Y)$ is a linear space over the field $\mathcal{C}$. If, moreover, $\mathcal{D}(A) = D$ and $\mathcal{R}(A) \subseteq D \subseteq Y$, then we write $\mathcal{F}(D)$ instead of $\mathcal{F}(D, D)$ or $\mathcal{F}(D, Y)$. However, it should be noticed that in this case if D is not a linear subspace of Y, then $\mathcal{F}(D)$ is no longer a linear space. Nevertheless, for any $A, B \in \mathcal{F}(D)$, we have $A \circ B \in \mathcal{F}(D)$.

For each $A \in \mathcal{F}(D)$, define $A^0 = I$, the identity operator, and $A^n = A \circ A^{n-1}$ inductively for $n = 1, 2, \dots$. It is evident that $A^n \in \mathcal{F}(D)$ for all n. By definition, for any $A \in \mathcal{F}(D), \mathcal{D}(I + A) = \mathcal{D}(I) \cap \mathcal{D}(A) = D$. Hence, if moreover $\mathcal{R}(I + A) \subseteq D$ (for example, when D is a linear subspace of Y), then we have $I + A \in \mathcal{F}(D)$ as well. Consequently, all the following composite operators

$$I, \ A, \ I + A, \ I + (I + A) \circ A, \ I + (I + (I + A) \circ A) \circ A, \ \dots \quad (2.6)$$

belong to $\mathcal{F}(D)$. In particular, for a linear operator A, if $\mathcal{R}(I+A) \subseteq D$ then

$$I, \ A, \ I + A, \ I + A + A^2, \ I + A + A^2 + A^3, \ \dots \quad (2.7)$$

are all in the same family $\mathcal{F}(D)$. In the linear case, $\mathcal{F}(D)$ is a *ring* with unity. But in the nonlinear case, $\mathcal{F}(D)$ is only a *near-ring* with unity in the sense that it has all properties of a ring with unity, such as operations of addition, scalar multiplication, and multiplication (i.e., composition), except the left distributiveness of multiplication over addition; namely, the equality $A \circ (B + C) = A \circ B + A \circ C$ for $A, B, C \in \mathcal{F}(D)$ does not hold in general for nonlinear operators.

§ 2. Lipschitz Operators

In this section, we let X and Y be two normed linear spaces over the field $\mathcal{C}$ of complex numbers. Let D be a subset of X and $\mathcal{F}(D, Y)$ be the family of operators A in $\mathcal{N}(X, Y)$ with $\mathcal{D}(A) = D$. Introduce a (semi)-norm into (a subset of) $\mathcal{F}(D, Y)$ by

$$\|A\| := \sup_{\substack{x_1, x_2 \in D \\ x_1 \neq x_2}} \frac{\|A(x_1) - A(x_2)\|_Y}{\|x_1 - x_2\|_X} \tag{2.8}$$

if it is finite. Note that although D is not necessarily a subspace of X, so that $x_1 - x_2$ may not be in D, $\|x_1 - x_2\|_X$ is well defined in (2.8) since X is a normed linear space. A simple example is the constant function $A(x) = c$ defined on $D = [0, \infty) \subset X = (-\infty, \infty)$, which is Lipschitz on D although $x_1, x_2 \in D$ may lead to $x_1 - x_2 \in (-\infty, \infty)$ in the calculation of its norm. Note also that in case A is linear and $D = X$, this defines a norm. In general, however, it is a semi-norm in the sense that $\|A\| = 0$ does not necessarily imply $A = 0$. In fact, it can be easily seen that $\|A\| = 0$ if and only if A is a constant-operator (need not be zero) that maps all elements from D to the same element in Y.

2.1. Definitions. Let $Lip(D, Y)$ be the subset of $\mathcal{F}(D, Y)$ with all its elements A satisfying $\|A\| < \infty$. Each $A \in Lip(D, Y)$ is called a *Lipschitz operator* mapping from D to Y, and the number $\|A\|$ is called the *Lipschitz semi-norm* of the operator A on D.

It is clear that an element A of $\mathcal{F}(D, Y)$ is in $Lip(D, Y)$ if and only if there is a number $L \geq 0$ such that

$$\|A(x_1) - A(x_2)\|_Y \leq L\|x_1 - x_2\|_X$$

for all $x_1, x_2 \in D$. Moreover, $\|A\|$ is the least such numbers L. It is also evident that *a Lipschitz operator is both bounded and continuous on its domain.*

We remark that under certain conditions, many nonlinear operators such as polynomic operators (of finite degrees), exponential operators, and analytic mappings with convergent Volterra series representations studied in the last chapter are of Lipschitz type. This observation is important from the application point of view, as will be illustrated in more detail later in the study of optimal feedback design strategies for general nonlinear control systems.

Lipschitz operators enjoy the following properties:

2.2. Theorem. *The class* $Lip(D, Y)$ *defined in 2.1 is a linear space over the field* C. *Moreover, if* $A, B \in Lip(D, Y)$ *and* $a \in C$, *then*

    (i) $\|A\| = 0$ *if and only if* A *is a constant operator on* D;

    (ii) $\|A + B\| \le \|A\| + \|B\|$; *and*

    (iii) $\|aA\| \le |a| \, \|A\|$.

The proof of this theorem is routine and is, hence, omitted.

As is well known, it is not easy to define an actual norm for a general nonlinear operator. For a Lipschitz operator, however, we have the following:

2.3. Theorem. *For any fixed* $x_0 \in D$, *the number*

$$\|A\|_{Lip} := \|A(x_0)\|_Y + \|A\|$$

$$= \|A(x_0)\|_Y + \sup_{\substack{x_1, x_2 \in D \\ x_1 \ne x_2}} \frac{\|A(x_1) - A(x_2)\|_Y}{\|x_1 - x_2\|_X}$$

defines a norm for all $A \in Lip(D, Y)$.

$\|A\|_{Lip}$ will be called the *Lipschitz norm* of A defined by $x_0 \in D$. A convenient choice for x_0 is of course $x_0 = 0$ if $0 \in D$, where note, however, that $A(0)$ is not zero in general if A is nonlinear. For a linear operator A, of course, the constant term $\|A(x_0)\|_Y$ can be dropped. In particular, if $x_0 = 0$ then actually $A(x_0) = 0$.

To prove the theorem, it amounts to showing that $\|A\|_{Lip} = 0$ implies $A = 0$, the zero operator. This, however, is an immediate consequence of part (i) of Theorem 2.2.

Now we can establish the following important result:

2.4. Theorem. *Let Y be a Banach space. Then, the family $Lip(D, Y)$ defined in 2.1 is a Banach space under the Lipschitz norm $\| \cdot \|_{Lip}$ defined in 2.3 by any fixed element $x_0 \in D$.*

Proof. Obviously, $Lip(D, Y)$ is a linear space since Y is a linear space and $\mathcal{D}(A) = D$ for all $A \in Lip(D, Y)$. Hence, it suffices to verify its completeness under the Lipschitz norm.

Let $\{A_n\}$ be a Cauchy sequence in $Lip(D, Y)$ such that $\|A_m - A_n\|_{Lip} \to 0$ as $m, n \to \infty$. Then, for any $x \in D$ we have

$$\|A_m(x) - A_n(x)\|_Y$$
$$\leq \ \|(A_m - A_n)(x) - (A_m - A_n)(x_0)\|_Y + \|A_m(x_0) - A_n(x_0)\|_Y$$
$$\leq \ \|A_m - A_n\|_{Lip}\|x - x_0\|_X + \|A_m(x_0) - A_n(x_0)\|_Y \, ,$$

which shows that the sequence $\{A_n\}$ is in fact uniformly Cauchy on each bounded subset of D. Since Y is complete, $A(x)$ exists and so is unique. Moreover, since $\{A_n\}$ is a Cauchy sequence, $\lim_{n\to\infty} \|A_n\|_{Lip} = c$, a constant, so that

$$\|A(x_1) - A(x_2)\|_Y = \lim_{n\to\infty} \|A_n(x_1) - A_n(x_2)\|_Y$$
$$\leq \lim_{n\to\infty} \|A_n\|_{Lip}\|x_1 - x_2\|_X$$
$$= c\|x_1 - x_2\|_X$$

for all $x_1, x_2 \in D$. This shows that $A \in Lip(D, Y)$ with

$$\|A\|_{Lip} \leq c + \|A(x_0)\|_Y \, .$$

We finally verify that $\|A_n - A\|_{Lip} \to 0$ as $n \to \infty$. Since the above also proves $\|A_n(x_0) - A(x_0)\|_Y \to 0$ as $n \to \infty$, for $\epsilon > 0$ we can let N be such that $\|A_m - A_n\|_{Lip} \leq \epsilon/2$ and $\|(A_n - A)(x_0)\|_Y \leq \epsilon/2$ for $m, n \geq N$. Then, for any $x_1, x_2 \in D$, it follows that

$$\|(A - A_n)(x_1) - (A - A_n)(x_2)\|_Y$$
$$= \lim_{m\to\infty} \|(A_m - A_n)(x_1) - (A_m - A_n)(x_2)\|_Y$$
$$\leq \lim_{m\to\infty} \|A_m - A_n\|_{Lip}\|x_1 - x_2\|_X$$
$$\leq \frac{\epsilon}{2}\|x_1 - x_2\|_X \, ,$$

so that $\|A_n - A\|_{Lip} \leq \epsilon$ for $n \geq N$. This implies that $\|A_n - A\|_{Lip} \rightarrow 0$ as $n \rightarrow \infty$, completing the proof of the theorem. $\qquad\square$

We remark that Theorem 2.4 has some useful consequences. For example, let us consider the following situation: Let X be a normed linear space and Y and Z be two Banach spaces. Let D be a subset of X. Denote by $Lip(D, Y)$ the family of Lipschitz operators A mapping from D to Y with $\mathcal{D}(A) = D$ and $\mathcal{R}(A)$ being a subspace of Y, and let $C : \mathcal{R}(A) \rightarrow Z$ be a bounded nonlinear operator with $\mathcal{D}(C) = \mathcal{R}(A), \mathcal{R}(C) \subseteq Z$, and with an arbitrary but well-defined operator norm $\|C\| < \infty$, such as the Lipschitz norm defined above, or the following norm:

$$\|C\| := \|C(0)\|_Z + \sup_{\substack{x \in \mathcal{D}(C) \\ x \neq 0}} \frac{\|C(x) - C(0)\|_Z}{\|x\|_Y}.$$

Moreover, set

$$S = \{\, A \in Lip(D, Y) : \quad C \circ A \in Lip(D, Z) \,\}. \qquad (2.9)$$

Then, we have the following result.

2.5. Corollary. *The family S of Lipschitz operators defined above is a Banach space.*

Since we have shown in Theorem 2.4 that $Lip(D, Y)$ is a Banach space, it is sufficient to note that S is a linear subspace of $Lip(D, Y)$. This is clear since all the operators in S have the same domain and Y is a linear space.

We next return to the near-ring $\mathcal{F}(D)$ studied in Section 1 above. Let X be a normed linear space (need not be a Banach space) and let D be a subset of X. Denote by $Lip(D)$ the family $Lip(D, D)$ of Lipschitz operators from D (the domain) to itself. Then, of course, $Lip(D) \subset \mathcal{F}(D)$. Moreover, we have the following:

2.6. Theorem. *If $A, B \in Lip(D)$, then $A \circ B \in Lip(D)$ with $\|A \circ B\|_{Lip} \leq \|A\|_{Lip}\|B\|_{Lip}$. Moreover, the identity operator I belongs to $Lip(D)$ with $\|I\|_{Lip} = 1$.*

Hence, $Lip(D)$ is itself a near-ring with unity I.

Proof. If $A, B \in Lip(D)$, with the Lipschitz norm $\| \cdot \|_{Lip}$ defined by 2.3, then we have

$$
\begin{aligned}
\|A \circ B(x_1) &- A \circ B(x_2)\|_X \\
&= \|A(B(x_1)) - A(B(x_2))\|_X \\
&\leq \|A\|\|B(x_1) - B(x_2)\|_X \\
&\leq \|A\| \, \|B\| \, \|x_1 - x_2\|_X \, ,
\end{aligned}
$$

for all $x_1, x_2 \in D$, where $\| \cdot \|$ is the Lipschitz semi-norm defined in Eq. (2.8). This implies that $A \circ B \in Lip(D)$ with $\|A \circ B\|_{Lip} \leq \|A\|_{Lip}\|B\|_{Lip}$. The second part of the theorem is obvious. $\qquad\square$

We are now concerned with inverse operators. Again, let X be a normed linear space and let D be a subset of X. Moreover, let $Lip(D)$ be defined as above.

2.7. Definitions. An element $A \in Lip(D)$ is said to be *invertible* in $Lip(D)$ if there exists a $B \in Lip(D)$ such that $A \circ B = B \circ A = I$, the identity operator. B is called the *inverse* of A and is denoted by A^{-1}.

The following results are evident from the above definition:

2.8. Theorem. *$A \in Lip(D)$ is invertible if and only if A is injective in the sense that $A(x_1) = A(x_2)$ implies $x_1 = x_2$, A is surjective in the sense that $\mathcal{R}(A) = D$, and the inverse mapping of A is in $Lip(D)$.*

2.9. Theorem. *If $A, B \in Lip(D)$ are both invertible in $Lip(D)$, then so is $A \circ B$. Moreover, $(A \circ B)^{-1} = B^{-1} \circ A^{-1}$. In general, if $A \in Lip(D)$ is invertible in $Lip(D)$, then so is A^n, where $A^n := A \circ A^{n-1}$, for all $n = 1, 2, \ldots$, with $A^0 = I$ and $(A^n)^{-1} = (A^{-1})^n$.*

A fundamental result on the invertibility of an element of $Lip(D)$, when D is a linear subspace, is given in the following theorem.

2.10. Theorem. *Let X be a Banach space and D be a linear subspace of X. If $A \in Lip(D)$ with $\|A\| < 1$, where $\|A\|$ is defined as in (2.8), then $I - A$ is invertible in $Lip(D)$ with*

$$\|(I - A)^{-1}\|_{Lip} \leq \|(I - A)^{-1}(x_0)\|_X + (1 - \|A\|)^{-1}$$

for any $x_0 \in \mathcal{R}(I-A)$. Moreover, if $B_0 := I$ and $B_n := I + A \circ B_{n-1}$ inductively for $n = 1, 2, \ldots$, then we have

$$\lim_{n \to \infty} B_n(x) = (I - A)^{-1}(x) \quad \text{for all } x \in D,$$

and

$$\|(I - A)^{-1}(x) - B_n(x)\|_X \leq \frac{\|A\|^n \|A(x)\|_X}{1 - \|A\|}$$

for all $x \in D$, $n = 0, 1, 2, \ldots$. Consequently, if $\|A\|_{Lip} < 1$, then

$$\|(I - A)^{-1}(x) - B_n(x)\|_X \leq \frac{\|A\|^n \|A(x)\|_X}{1 - \|A\|} \leq \frac{\|A\|_{Lip}^n \|A(x)\|_X}{1 - \|A\|_{Lip}}$$

for all $x \in D$, $n = 0, 1, 2, \ldots$.

Proof. First, observe that since $A \in Lip(D)$ and D is a linear subspace, we have $I - A \in Lip(D)$. Hence, for any $x_1, x_2 \in D$,

$$\|(I - A)(x_1) - (I - A)(x_2)\|_X$$
$$\geq \|x_1 - x_2\|_X - \|A(x_1) - A(x_2)\|_X$$
$$\geq (1 - \|A\|)\|x_1 - x_2\|_X,$$

which implies that $I - A$ is injective. By Theorem 2.8, we have to show that $I - A$ is surjective and $(I - A)^{-1} \in Lip(D)$.

We first prove that $I - A$ is surjective. For any $x \in D$, we will show that

$$\|B_{n+1}(x) - B_n(x)\|_X \leq \|A\|^n \|A(x)\|_X, \quad n = 0, 1, 2, \ldots.$$

For $n = 0$, it is obviously true. Suppose that this inequality holds for $n = k - 1$. Then, since

$$\|B_{k+1}(x) - B_k(x)\|_X$$
$$= \|A \circ B_k(x) - A \circ B_{k-1}(x)\|_X$$
$$\leq \|A\| \, \|B_k(x) - B_{k-1}(x)\|_X$$
$$\leq \|A\| \, \|A\|^{k-1} \|A(x)\|_X,$$

the inequality is true for all $n, n = 0, 1, 2, \ldots$. Consequently, for any positive integer m, we have

$$
\begin{aligned}
\|B_{n+m}(x) &- B_n(x)\|_X \\
&= \left\| \sum_{k=0}^{m-1} \left(B_{n+k+1}(x) - B_{n+k}(x) \right) \right\|_X \\
&\leq \sum_{k=0}^{m-1} \|B_{n+k+1}(x) - B_{n+k}(x)\|_X \\
&\leq \sum_{k=0}^{m-1} \|A\|^{n+k} \|A(x)\|_X \\
&\leq \frac{\|A\|^n \|A(x)\|_X}{1 - \|A\|} .
\end{aligned}
$$

Since $\|A\| < 1$ and X is a Banach space, this implies that $\{B_n(x)\}$ is a Cauchy sequence in D, so that $\lim_{n \to \infty} B_n(x) = C(x)$ in norm for some operator C for any $x \in D$. We next show that $C = (I - A)^{-1} \in Lip(D)$. In doing so, observe that for any $x \in D$,

$$
\begin{aligned}
\|C(x) - B_n(x)\|_X &= \lim_{m \to \infty} \|B_{n+m}(x) - B_n(x)\|_X \\
&\leq \frac{\|A\|^n \|A(x)\|_X}{1 - \|A\|}
\end{aligned}
$$

and recall that A is Lipschitz and hence is continuous. Consequently, we have

$$
\begin{aligned}
C(x) &= \lim_{n \to \infty} B_n(x) \\
&= \lim_{n \to \infty} (I + A \circ B_{n-1})(x) \\
&= x + A \circ C(x), \quad x \in D,
\end{aligned}
$$

where again the convergence is in norm. It then follows that $C = I + A \circ C$, so that $(I - A) \circ C = I$, or C is a right inverse of $I - A$. Note, furthermore, that this also implies $(I - A)$ is surjective. Hence, $(I - A)^{-1}$ exists on D.

Next, for any $y_1, y_2 \in D$, let $x_1, x_2 \in D$ with $y_1 = (I - A)(x_1)$ and $y_2 = (I - A)(x_2)$. Then since $\|A\| < 1$, it follows, from the inequality shown at the beginning of the proof, that

$$
\begin{aligned}
\|(I - A)^{-1}(y_1) &- (I - A)^{-1}(y_2)\|_X \\
&= \|x_1 - x_2\|_X \\
&\leq (1 - \|A\|)^{-1}\|(I - A)(x_1) - (I - A)(x_2)\|_X \\
&= (1 - \|A\|)^{-1}\|y_1 - y_2\|_X \,,
\end{aligned}
$$

which implies that the inverse mapping of $I - A$ is in $Lip(D)$ with $\|(I - A)^{-1}\| \leq (1 - \|A\|)^{-1}$, so that

$$
\|(I - A)^{-1}\|_{Lip} \leq \|(I - A)^{-1}(x_0)\|_X + (1 - \|A\|)^{-1} \,.
$$

In summary, we have $C = (I - A)^{-1} \in Lip(D)$, completing the proof of the first part of the theorem. The second part of the theorem has actually been verified in the above proof. $\qquad\square$

We remark that for the special case where $A \in Lip(D)$ is a linear Lipschitz operator, we have the following:

2.11. Corollary. *Let X be a Banach space and D be a linear subspace of X. Let $A \in Lip(D)$ be linear with $\|A\| < 1$. Then $I - A$ is invertible in $Lip(D)$ with*

$$
\|(I - A)^{-1}\| \leq (1 - \|A\|)^{-1} \,.
$$

Moreover,

$$
(I - A)^{-1} = \sum_{n=0}^{\infty} A^n \,,
$$

where the infinite sum is absolutely convergent, with

$$
\left\| (I - A)^{-1} - \sum_{k=0}^{n} A^k \right\| \leq \frac{\|A\|^{n+1}}{1 - \|A\|}
$$

for all $n = 0, 1, 2, \ldots$.

Finally, let us consider a sum of two nonlinear Lipschitz operators.

2.12. Theorem. *Let X be a Banach space and D be a linear subspace of X. Let $A, B \in Lip(D)$ and suppose that A is invertible in $Lip(D)$ with $\|B\| \, \|A^{-1}\| < 1$. Then, $A+B$ is invertible in $Lip(D)$ with*

$$\|(A + B)^{-1}\|_{Lip}$$
$$\leq \|A^{-1}\| \big(\|(A + B)^{-1}(x_0)\|_X + (1 - \|B\| \, \|A^{-1}\|)^{-1}\big)$$

for any $x_0 \in D$. In particular, the set of all invertible operators in $Lip(D)$ is open in the Banach space $Lip(D)$. If, furthermore, $\|B\|_{Lip}\|A^{-1}\|_{Lip} < 1$, then $A + B$ is invertible in $Lip(D)$ with

$$\|(A + B)^{-1}\|_{Lip}$$
$$\leq \|A^{-1}\|_{Lip}\big(\|(A + B)^{-1}(x_0)\|_X + (1 - \|B\|_{Lip}\|A^{-1}\|_{Lip})^{-1}\big)$$

for any $x_0 \in D$.

Proof. Since $A + B = (I + B \circ A^{-1}) \circ A$ and

$$\|B \circ A^{-1}\| \leq \|B\| \, \|A^{-1}\| < 1\,,$$

it follows from Theorem 2.10 that the inverse operator $(I + B \circ A^{-1})^{-1}$ exists in $Lip(D)$ with

$$\|(I + B \circ A^{-1})^{-1}\|_{Lip} \leq \|(A + B)^{-1}(0)\|_X + (1 - \|B\| \, \|A^{-1}\|)^{-1}\,.$$

Consequently,

$$(A + B)^{-1} = A^{-1} \circ (I + B \circ A^{-1})^{-1}\,,$$

with

$$\|(A + B)^{-1}\|_{Lip}$$
$$\leq \|A^{-1}\| \, \|(I + B \circ A^{-1})^{-1}\|_{Lip}$$
$$\leq \|A^{-1}\|\big(\|(A + B)^{-1}(0)\|_X + (1 - \|B\| \, \|A^{-1}\|)^{-1}\big)\,.$$

The last inequality is obvious. $\qquad\qquad\qquad\qquad\qquad\qquad\square$

§ 3. Contraction Mapping Theorem

In this section, we study the important contraction mapping theorem in an extended form for Lipschitz operators and some of its consequences.

2.13. Definition. Let X be a normed linear space and let D be a subset of X. A Lipschitz operator $A : D \to X$ is called a *contraction on D* if its semi-norm $\|A\| < 1$ on D.

Before we state and prove the contraction mapping theorem, we need the following lemma, which is valid for a general mapping (need not be Lipschitz) defined on any nonempty set.

2.14. Lemma. *Let D be a nonempty set in X and let $A : D \to D$ be a nonlinear mapping. For $k = 1, 2, \ldots$, define*

$$S_k = \{ \, x \in D : \; A^k(x) = x \, \}.$$

If for some $n \geq 1$ we have $S_n = \{x_0\}$, a singleton in D, then $S_1 = \{x_0\}$.

Proof. Since $A^n(x_0) = x_0$, we have

$$A^n\big(A(x_0)\big) = A\big(A^n(x_0)\big) = A(x_0),$$

which implies that $A(x_0) \in S_n$. Since $S_n = \{x_0\}$, a singleton, we have $A(x_0) = x_0$. Hence, $x_0 \in S_1$, namely: $S_1 \neq \emptyset$.
  Suppose that $y \in S_1$. Then, $A(y) = y$, so that $A\big(A(y)\big) = A(y) = y$, or

$$y = A(y) = A^2(y) = \cdots = A^n(y).$$

This implies that $y \in S_n = \{x_0\}$. Hence $y = x_0$, or $S_1 = \{x_0\}$, completing the proof of the lemma. $\square$

We now establish an extended version of the contraction mapping theorem relative to Lipschitz operators.

2.15. Theorem. *Let X be a Banach space and let $A : X \to X$ be an operator that is not necessarily Lipschitz. For a fixed $x_0 \in X$ and $\rho > 0$, set*

$$D = \{\, x \in X : \ \|x - x_0\|_X \le \rho \,\}.$$

Assume that there exists an integer $n \ge 1$ such that A^n is Lipschitz with $L = \|A^n\| < 1$ on D and such that $\|A^n(x_0) - x_0\|_X \le (1 - L)\rho$. Then, there exists a unique element $x^* \in D$ such that $A(x^*) = x^*$. Moreover, if for this integer n we have

$$x_k = A^n(x_{k-1}), \qquad k = 1, 2, \ldots,$$

then $x_k \in D$ with

$$\|x_k - x^*\|_X \le \frac{L^k}{1 - L}\|x_0 - A^n(x_0)\|_X \le L^k \rho.$$

Consequently, $\|x_k - x^*\|_X \to 0$ as $k \to \infty$.

The point x^* described above is called a *fixed point* of the operator A in D.

Proof. We first show that $x_k \in D$ for all $k = 1, 2, \ldots$. Since $x_0 \in D$, it follows from the assumption that

$$\|x_1 - x_0\|_X = \|A^n(x_0) - x_0\|_X \le (1 - L)\rho \le \rho,$$

so that $x_1 \in D$. By induction, suppose that $x_m \in D$ for $m \le k - 1$. Then, for any $i \ge 1$, we have

$$\begin{aligned}
\|x_i - x_{i-1}\|_X &= \|A^n(x_{i-1}) - A^n(x_{i-2})\|_X \\
&\le L\|x_{i-1} - x_{i-2}\|_X \\
&\le \cdots \\
&\le L^{i-1}\|x_1 - x_0\|_X \\
&\le L^{i-1}\|A^n(x_0) - x_0\|_X,
\end{aligned}$$

so that

$$\|x_i - x_{i-1}\|_X \le L^{i-1}(1 - L)\rho, \quad i \ge 1.$$

It follows that

$$\|x_k - x_0\| \le \|x_k - x_{k-1}\|_X + \cdots + \|x_1 - x_0\|_X$$

$$\le \sum_{i=1}^{k} L^{i-1}(1-L)\rho$$

$$= (1 - L^k)\rho$$

$$\le \rho,$$

that is, $x_k \in D$ as claimed. This completes the induction.

Next, we show that $\{x_k\}$ is a Cauchy sequence in D. Pick any $m, k \ge 1$. Then, it follows from the above proof that

$$\|x_{k+m} - x_k\|_X \le \|x_{k+m} - x_{k+m-1}\|_X + \cdots + \|x_{k+1} - x_k\|_X$$

$$\le \left(L^{k+m-1} + \cdots + L^k\right)\|A^n(x_0) - x_0\|_X$$

$$= \frac{(L^k - L^{k+m})}{1 - L}\|A^n(x_0) - x_0\|_X,$$

which shows that $\{x_k\}$ is a Cauchy sequence. Consequently, $x_k \to x^*$ for some $x^* \in D$ as $k \to \infty$, since D is closed in the Banach space X.

We then show that x^* is the unique element in D such that $A(x^*) = x^*$.

Since A^n is Lipschitz and hence is continuous on D, we have

$$\|A^n(x_k) - A^n(x^*)\|_X \to 0 \quad \text{or} \quad \|x_{k+1} - A^n(x^*)\|_X \to 0$$

as $k \to \infty$. This implies that $x^* = A^n(x^*)$. To show the uniqueness of x^*, assume that $A^n(y) = y$ for some $y \in D$. Then,

$$\|x^* - y\|_X = \|A^n(x^*) - A^n(y)\|_X \le L\|x^* - y\|_X,$$

which implies that $(1 - L)\|x^* - y\|_X \le 0$, or $x^* = y$. Hence, by Lemma 2.14, x^* is the unique element in D satisfying $A(x^*) = x^*$.

Finally, letting $m \to \infty$ in the inequality

$$\|x_{k+m} - x_k\|_X \le \frac{L^k - L^{k+m}}{1 - L}\|A^n(x_0) - x_0\|_X$$

obtained above, we arrive at

$$\|x^* - x_k\|_X \le \frac{L^k}{1 - L}\|A^n(x_0) - x_0\|_X \le L^k \rho,$$

as claimed. $\qquad\qquad\qquad\qquad\qquad\qquad\qquad\qquad\qquad\square$

2.16. Corollary. *Let X be a Banach space and let $A : X \to X$ be an operator (not necessarily Lipschitz). If there exists an $n \geq 1$ such that A^n is Lipschitz on X with $L = \|A^n\| < 1$, then there exists a unique element $x^* \in X$ such that $A(x^*) = x^*$. Moreover, if $x_0 \in X$ and*

$$x_k = A^n(x_{k-1}), \quad k = 1, 2, \dots,$$

then $\|x_k - x^*\|_X \to 0$ as $k \to \infty$, with

$$\|x_k - x^*\|_X \leq \frac{L^k}{1 - L} \|x_0 - A^n(x_0)\|_X.$$

The point x^* described above is called a *fixed point* of the operator A in X.

Proof. We apply Theorem 2.15. Let $x_0 \in X$ be arbitrarily chosen. Pick

$$\rho \geq \frac{1}{1 - L} \|x_0 - A^n(x_0)\|_X$$

and define

$$D = \{ x \in X : \quad \|x - x_0\|_X \leq \rho \}.$$

Denote by L_D the Lipschitz semi-norm of A^n on D. Then, $L_D \leq L$. Hence, $(1 - L_D)^{-1} < (1 - L)^{-1}$, so that

$$\frac{1}{1 - L_D} \|x_0 - A^n(x_0)\|_X \leq \frac{1}{1 - L} \|x_0 - A^n(x_0)\|_X \leq \rho.$$

Consequently, $\|x_0 - A^n(x_0)\|_X \leq (1 - L_D)\rho$, that is, the conditions stated in Theorem 2.15 are satisfied. Thus, there exists a unique element $x^* \in D \subset X$ such that $A(x^*) = x^*$. The rest can be easily verified. $\square$

As an application of the contraction mapping theorem, we show the invertibility of a nonlinear integral operator in the following example.

2.17. Example. Let $T \in (0, \infty)$ and set

$$\Omega = \{ (t, s) \in R^2 : \quad 0 \le s \le t \le T \}.$$

Let $K : \Omega \times R^n \times R^n \to R^n$ be a continuous function such that

$$\|K(t, s, x_1, y_1) - K(t, s, x_2, y_2)\| \le \alpha \|x_1 - x_2\| + \beta \|y_1 - y_2\|$$

for some $\alpha, \beta > 0$, all $(t, s) \in \Omega$, and all $x_1, x_2, y_1, y_2 \in R^n$. Here, $\|\cdot\|$ denotes the usual Euclidean norm in R^n. Furthermore, for any $\sigma \in [0, T]$, let $g : [-\sigma, 0] \to R^n$ be a continuous function, and for any continuous function $f(t) \in C^n[0, T]$, the n-dimensional space of continuous functions defined on $[0, T]$, define a "σ-function" as follows:

$$f_\sigma(t) = \begin{cases} f(t - \sigma) & \sigma \le t \le T \\ g(t - \sigma) & 0 \le t < \sigma, \end{cases}$$

which is a piecewise continuous function with a possible jump at $t = \sigma$. Then, for a fixed $\sigma \in [0, T]$ and for given continuous functions K and g as described above, define a nonlinear integral operator A on $C^n[0, T]$ by

$$A(f)(t) = f(t) + \int_0^t K\big(t, s, f(s), f_\sigma(s)\big)ds, \quad 0 \le t \le T. \quad (2.10)$$

Obviously, A maps $C^n[0, T]$ into itself. We will show that A is invertible on $C^n[0, T]$.

To do so, choose arbitrarily an $h \in C^n[0, T]$, and consider a new operator B defined on $C^n[0, T]$ by

$$B(f)(t) = h(t) - \int_0^t K\big(t, s, f(s), f_\sigma(s)\big)ds, \quad 0 \le t \le T. \quad (2.11)$$

It is evident that B maps $C^n[0, T]$ into itself. It is also clear that for any $f_1, f_2 \in C^n[0, T]$, we have

$$\|(B(f_1) - B(f_2))(t)\|$$
$$\le \int_0^t \{\alpha \|f_1(s) - f_2(s)\| + \beta \|f_{1\sigma}(s) - f_{2\sigma}(s)\|\}ds$$
$$\le t(\alpha + \beta)\|f_1 - f_2\|_\infty,$$

where $\|f\|_\infty := \sup_{0 \le t \le T} \|f(t)\|$. In general, by induction we can easily show that

$$\|(B^k(f_1) - B^k(f_2))(t)\| \le \frac{t^k (\alpha + \beta)^k}{k!} \|f_1 - f_2\|_\infty. \qquad (2.12)$$

Indeed, if this is true for some $k \ge 1$, then

$$\|(B^{k+1}(f_1) - B^{k+1}(f_2)(t))\|$$
$$= \|(B(B^k(f_1)) - B(B^k(f_2)))(t)\| \le \lambda_1 + \lambda_2,$$

where

$$\lambda_1 = \int_0^t \alpha \|(B^k(f_1) - B^k(f_2))(s)\| ds$$

$$\lambda_2 = \int_0^t \beta \|((B^k(f_1))_\sigma - (B^k(f_2))_\sigma)(s)\| ds.$$

By the induction hypothesis, we have

$$\lambda_1 \le \frac{t^{k+1} \alpha (\alpha + \beta)^k}{(k+1)!} \|f_1 - f_2\|_\infty.$$

As to λ_2, we have $\lambda_2 = 0$ for $t < \sigma$ by definition of the σ-function. For $t \ge \sigma$, by a change of variables $\tau = s - \sigma$, we have

$$\lambda_2 = \int_\sigma^t \beta \|((B^k(f_1))_\sigma - (B^k(f_2))_\sigma)(s)\| ds$$

$$= \int_\sigma^t \beta \|(B^k(f_1) - B^k(f_2))(s - \sigma)\| ds$$

$$= \int_0^{t-\sigma} \beta \|(B^k(f_1) - B^k(f_2))(\tau)\| d\tau$$

$$\le \int_0^{t-\sigma} \beta \frac{\tau^k (\alpha + \beta)^k}{k!} \|f_1 - f_2\|_\infty d\tau$$

$$= \frac{(t-\sigma)^{k+1} \beta (\alpha + \beta)^k}{(k+1)!} \|f_1 - f_2\|_\infty$$

$$\le \frac{t^{k+1} \beta (\alpha + \beta)^k}{(k+1)!} \|f_1 - f_2\|_\infty.$$

Hence, for any $t \in [0, T]$, we have

$$\lambda_1 + \lambda_2 \le \frac{t^{k+1}(\alpha + \beta)^{k+1}}{(k+1)!} \|f_1 - f_2\|_\infty,$$

as claimed. This completes the induction.

Now, it follows from inequality (2.12) that

$$\|B^k(f_1) - B^k(f_2)\|_\infty \le \frac{T^k(\alpha + \beta)^k}{k!} \|f_1 - f_2\|_\infty.$$

Observe that for large values of k, we have $T^k(\alpha+\beta)^k/k! < 1$. This implies that the operator B satisfies the conditions of Corollary 2.16. Hence, there exists a unique $f^* \in C^n[0, T]$ such that $B(f^*) = f^*$. Consequently, Eqs. (2.10) and (2.11) together imply $A(f^*) = h$, namely, the operator A is invertible on $C^n[0, T]$.

As a special case of this example, the following nonlinear Volterra integral operator is invertible.

2.18. Example. Let $T \in (0, \infty)$ and set

$$\Omega = \{ (t, s) \in R^2 : \quad 0 \le s \le t \le T \}.$$

Let $K : \Omega \times R^n \to R^n$ be a continuous function such that

$$\|K(t, s, f_1(s)) - K(t, s, f_2(s))\| \le \alpha \|f_1(s) - f_2(s)\|$$

for some $\alpha > 0$, all $(t, s) \in \Omega$, and all $f \in R^n$. Here, again, $\|\cdot\|$ denotes the usual Euclidean norm in R^n. Then, the nonlinear Volterra integral operator V defined by

$$V(f) = f(t) + \int_0^t K(t, s, f(s))ds, \quad 0 \le t \le T, \qquad (2.13)$$

is invertible on $C^n[0, T]$, where $C^n[0, T]$ is the space of n-tuples of continuous functions defined on $[0, T]$.

We remark that in the above two examples, if the assumptions on the kernel function K are suitably modified, then analogous results hold for the more general case where the space $C^n[0, T]$ is replaced by the space of n-tuples of functions in $L_p^n[0, T]$ for any $1 \le p \le \infty$.

§ 4. Generalized Lipschitz Operators

We have studied in Section 2 the basic theory of nonlinear Lipschitz operators. In this section, we extend this notion to a more general setting that is suitable for nonlinear systems control theory and applications. More precisely, we will generalize the Lipschitz operator theory developed previously from a standard normed linear space setting to one defined under a so-called extended linear space framework, which will be seen in the next few chapters to be very useful for nonlinear systems control theory and engineering in the considerations of stability, causality, robustness, uniqueness of internal control signals, and coprime factorizations, as well as optimal compensators design.

To develop this general theory, we first need the notion of extended linear spaces. An extended normed linear space, or simply extended linear space, is not complete in norm in general (and hence is not a Banach space), but it is sometimes called an extended Banach space in the literature, indicating that it is determined by a relative Banach space.

Let M be the family of real-valued measurable functions defined on $[0, \infty)$, which is a linear space. For each constant $T \in [0, \infty)$, let P_T be the *projection operator* mapping from M to another linear space, M_T, of measurable functions such that

$$f_T(t) := P_T(f)(t) = \begin{cases} f(t) & t \leq T \\ 0 & t > T, \end{cases} \qquad (2.14)$$

where $f_T(t) \in M_T$ is called the *truncation* of $f(t)$ with respect to T. Then, for any given Banach space X of measurable functions, set

$$X^e = \left\{ f \in M : \ \|f_T\|_X < \infty \ \text{ for all } \ T < \infty \right\}. \qquad (2.15)$$

Obviously, X^e is a linear subspace of M. The space X^e so defined is called the *extended linear space associated with the Banach space* X.

Here, we note again that the space X^e is usually not complete in norm. In fact, an element in X^e only has the local norm-boundedness, so that even a norm cannot be well defined in X^e

in general. For example, let $X = L_\infty(0, \infty)$. Then, the function $f(t) = e^t$ is in $X^e = L_\infty^e(0, \infty)$ defined as above but is certainly not in $X = L_\infty(0, \infty)$.

One reason for considering extended linear spaces is that all control signals in real life are time-limited, but in the study of a control processing we sometimes do not know when the processing will stop. Hence, because of the finite time-duration in practice, functions $f(t) = e^t$ and $f(t) = t^2$ and the like should be accepted in the underlying spaces. Such functions are not in any of the standard Banach spaces, but they are in the extended linear spaces, as can be easily verified.

Another reason for considering extended linear spaces for nonlinear control systems is that if the norm in the Banach space X is suitably defined such that $\|f_T\|_X$ is monotonically increasing as $T \to \infty$ for all $f \in X^e$ and such that $\|f_T\|_X \to \|f\|_X$ as $T \to \infty$ for all $f \in X$, then many useful techniques and results can be carried over from the standard Banach space X to the extended linear space X^e. This viewpoint will be clear if we take the so-called *causality* into account, which is a basic requirement for a realizable physical control system. We will come back to this issue in the next chapter.

Next, we introduce the notion of generalized nonlinear Lipschitz operators defined on an extended linear space setting as follows.

2.19. Definition. Let X^e and Y^e be two extended linear spaces, which are associated respectively with two given Banach spaces X and Y of measurable functions defined on the time domain $[0, \infty)$, and let D be a subset of X^e. A nonlinear operator $A : D \to Y^e$ is called a *generalized Lipschitz operator* on D if there exists a constant L such that

$$\left\| [A(x)]_T - [A(\tilde{x})]_T \right\|_Y \le L \|x_T - \tilde{x}_T\|_X$$

for all $x, \tilde{x} \in D$ and for all $T \in [0, \infty)$.

Note that the least such constants L is given by

$$\|A\| := \sup_{T \in [0, \infty)} \sup_{\substack{x, \tilde{x} \in D \\ x_T \ne \tilde{x}_T}} \frac{\left\| [Ax]_T - [A\tilde{x}]_T \right\|_Y}{\|x_T - \tilde{x}_T\|_X}, \qquad (2.16)$$

which is a semi-norm for general nonlinear operators and is the actual norm for linear operators A. The actual norm for a nonlinear operator A is given by

$$\|A\|_{Lip} = \|A(x_0)\|_Y + \|A\| = \|A(x_0)\|_Y$$
$$+ \sup_{T\in[0,\infty)} \sup_{\substack{x,\tilde{x}\in D \\ x_T\neq\tilde{x}_T}} \frac{\|[A(x)]_T - [A(\tilde{x})]_T\|_Y}{\|x_T - \tilde{x}_T\|_X} \quad (2.17)$$

for any fixed $x_0 \in D$. This can be easily verified by definition. To compare them with the (standard) nonlinear Lipschitz operators defined on a Banach space, the reader is referred to Section 2 above, where X^e and Y^e are simply replaced by X and Y, respectively, and all subscripts T (for the truncations) are dropped.

Here, it follows immediately that for any $T \in [0,\infty)$, we have

$$\|[A(x)]_T - [A(\tilde{x})]_T\|_Y \leq \|A\| \|x_T - \tilde{x}_T\|_X \leq \|A\|_{Lip}\|x_T - \tilde{x}_T\|_X .$$

We remark that the family of standard Lipschitz operators and the family of generalized Lipschitz operators are not comparable since they have different domains and ranges. However, it can be easily verified that when the extended linear spaces become standard with all the subscripts T being dropped, generalized Lipschitz operators become standard ones. It can also be verified that many standard Lipschitz operators are also extended Lipschitz. For example, it is clear from Eq. (2.16) that all bounded linear operators are not only standard Lipschitz on Banach spaces but also generalized Lipschitz on extended linear spaces.

To see that the family of standard Lipschitz operators and the family of generalized Lipschitz operators are not the same, particularly that the latter may not contain the former, we show the following simple example.

2.20. Example. Let $x_1(t) = 0$ and

$$x_2(t) = \begin{cases} 0 & 0 \leq t \leq 1 \\ 1 & 1 < t < \infty, \end{cases}$$

and set $D = \{x_1(t), x_2(t)\}$. Then D is a subset of both $L_\infty(0, \infty)$ and $L_\infty^e(0, \infty)$. Consider the nonlinear operator $A : D \to L_\infty(0, \infty)$ defined by

$$A(x) = \begin{cases} 0 & \text{if } x = x_1 \\ 1 & \text{if } x = x_2 \,. \end{cases}$$

Then, since $\|x_2\|_{L_\infty} = 1$ but $\inf_{0 \leq T < \infty} \|[x_2]_T\|_{L_\infty} = 0$, so that

$$\frac{\|A(x_1) - A(x_2)\|_{L_\infty}}{\|x_1 - x_2\|_{L_\infty}} = \frac{\|1\|_{L_\infty}}{\|x_2\|_{L_\infty}} = 1$$

and

$$\sup_{0 \leq T < \infty} \frac{\|[A(x_1)]_T - [A(x_2)]_T\|_{L_\infty}}{\|[x_1]_T - [x_2]_T\|_{L_\infty}} = \sup_{0 \leq T < \infty} \frac{\|[1]_T\|_{L_\infty}}{\|[x_2]_T\|_{L_\infty}} = \infty \,,$$

the nonlinear operator A is a standard Lipschitz operator but not a generalized one on D.

For simplicity, throughout this section, by a *Lipschitz operator* we will always mean one defined in this generalized sense. The norms $\|\cdot\|$ and $\|\cdot\|_{Lip}$ used in this section are defined by Eqs. (2.16) and (2.17), respectively.

We then have the following result.

2.21. Theorem. *Let X^e and Y^e be two extended linear spaces associated with two given Banach spaces X and Y, respectively, and let D be a subset of X^e. Then, the following family of Lipschitz operators is a Banach space:*

$$Lip(D, Y^e) = \left\{ A : D \to Y^e : \|A\|_{Lip} < \infty \text{ on } D \right\}. \qquad (2.18)$$

The proof of the theorem is similar to, but different at the technical level from, that of Theorem 2.4.

Proof. First, it is clear that $Lip(D, Y^e)$ is a normed linear space. Hence, it is sufficient to verify its completeness.

Let $\{A_n\}$ be a Cauchy sequence in $Lip(D, Y^e)$ such that $\|A_m - A_n\|_{Lip} \to 0$ as $m, n \to \infty$. We need to show that $\|A_n - A\|_{Lip} \to 0$ for some $A \in Lip(D, Y^e)$ as $n \to \infty$.

Let $T \in [0, \infty)$ be fixed. For any $\tilde{x} \in D$, by definition of the Lipschitz norm with an $x_0 \in D$, we have

$$\left\| [(A_m - A_n)(\tilde{x})]_T - [(A_m - A_n)(x_0)]_T \right\|_X$$
$$\leq \|A_m - A_n\|_{Lip} \|\tilde{x}_T - [x_0]_T\|_X ,$$

so that

$$\left\| [A_m(\tilde{x})]_T - [A_n(\tilde{x})]_T \right\|_X$$
$$= \left\| [(A_m - A_n)(\tilde{x})]_T \right\|_X$$
$$\leq \left\| [(A_m - A_n)(x_0)]_T \right\|_X + \|A_m - A_n\|_{Lip} \|\tilde{x}_T - [x_0]_T\|_X .$$

Since the right-hand side of the above tends to zero as $m, n \to \infty$, it follows that the sequence $\{[A_n(\tilde{x})]_T\}$ is Cauchy in the range Y (and in fact is uniformly Cauchy over each bounded subset of the domain D). Hence, for each fixed $T \in [0, \infty)$, $\tilde{v}_T := \lim[A_n(\tilde{x})]_T$ exists in the range Y (and is uniform over bounded subsets of the domain D). Let v be a function such that $v_T = \tilde{v}_T$ for all $T \in [0, \infty)$, and define a nonlinear operator A by $A : \tilde{x} \to v$. Then, A satisfies $[A(\tilde{x})]_T = \tilde{v}_T$ for all $T \in [0, \infty)$. We will show that $A \in Lip(D, Y^e)$. We first note that the operator A so defined has domain D since in the above $\tilde{x} \in D$ is arbitrary. We then observe that A is actually independent of T. Then, since $\|A_m - A_n\|_{Lip} \to 0$ we have $\lim \|A_n\|_{Lip} = c$, a constant, so that for any $\tilde{x}_1, \tilde{x}_2 \in D$,

$$\left\| [A(\tilde{x}_1)]_T - [A(\tilde{x}_2)]_T \right\|_X = \lim_{n \to \infty} \left\| [A_n(\tilde{x}_1)]_T - [A_n(\tilde{x}_2)]_T \right\|_X$$
$$\leq \lim_{n \to \infty} \|A_n\|_{Lip} \|[\tilde{x}_1]_T - [\tilde{x}_2]_T\|_X$$
$$= c\|[\tilde{x}_1]_T - [\tilde{x}_2]_T\|_X .$$

Therefore, taking the supremum over D and then the supremum over $[0, \infty)$ yields

$$\sup_{T \in [0,\infty)} \sup_{\substack{\tilde{x}_1, \tilde{x}_2 \in D \\ [\tilde{x}_1]_T \neq [\tilde{x}_2]_T}} \frac{\left\| [A(\tilde{x}_1)]_T - [A(\tilde{x}_2)]_T \right\|_X}{\|[\tilde{x}_1]_T - [\tilde{x}_2]_T\|_X} \leq c,$$

which implies that $\|A\| \leq c < \infty$, so that $A \in Lip(D, Y^e)$.

We finally verify that $\|A_n - A\|_{Lip} \to 0$ as $n \to \infty$. Since the above also proves (letting $\tilde{x} = x_0$ therein) that $\|[A_n(x_0)]_T - [A(x_0)]_T\|_X \to 0$ as $n \to \infty$ for each $T \in [0, \infty)$, for $\epsilon > 0$ we can let N be such that $\|A_m - A_n\|_{Lip} \leq \epsilon/2$ and $\|[(A_n - A)(x_0)]_T\|_X \leq \epsilon/2$ for $m, n \geq N$. Then for any $\tilde{x}_1, \tilde{x}_2 \in D$, we have

$$\|[(A - A_n)(\tilde{x}_1)]_T - [(A - A_n)(\tilde{x}_2)]_T\|_X$$
$$= \lim_{k \to \infty} \|[(A_k - A_n)(\tilde{x}_1)]_T - [(A_k - A_n)(\tilde{x}_2)]_T\|_X$$
$$\leq \lim_{k \to \infty} \|A_k - A_n\|_{Lip} \|[\tilde{x}_1]_T - [\tilde{x}_2]_T\|_X$$
$$\leq \frac{\epsilon}{2} \|[\tilde{x}_1]_T - [\tilde{x}_2]_T\|_X \,,$$

so that $\|A_n - A\|_{Lip} \leq \epsilon$ for $n \geq N$. This shows that $\|A_n - A\|_{Lip} \to 0$ as $n \to \infty$ and completes the proof of the theorem. $\qquad \square$

The following important result is a generalized counterpart of Theorem 2.10. Here, $Lip(D) = Lip(D, D)$, as defined in (2.18) with $Y^e = D$.

2.22. Theorem. *Let D be a linear subspace of the extended linear space X^e associated with a given Banach space X, and let $A \in Lip(D)$ with $\|A\| < 1$, where $\|\cdot\|$ is the semi-norm for the Lipschitz operator as defined in (2.16). Then, the operator $I - A$ is invertible on D with*

$$\|(I - A)^{-1}\|_{Lip} \leq \|(I - A)^{-1}(x_0)\|_X + (1 - \|A\|)^{-1}$$

for any $x_0 \in D$. Moreover, if $C_0 := I$ and $C_n := I + A \circ C_{n-1}$ inductively for $n = 1, 2, \ldots$, then for each fixed $T \in [0, \infty)$ and for all $\tilde{x} \in D$, we have

$$\lim_{n \to \infty} [C_n(\tilde{x})]_T = [(I - A)^{-1}(\tilde{x})]_T \,,$$

with the error bound

$$\|[(I - A)^{-1}(\tilde{x})]_T - [C_n(\tilde{x})]_T\|_X \leq \frac{\|A\|^n \|[A(\tilde{x})]_T\|_X}{1 - \|A\|} \,,$$

$n = 0, 1, 2, \ldots$. *Consequently, if* $\|A\|_{Lip} < 1$, *then for each* $T \in [0, \infty)$ *and for all* $\tilde{x} \in D$, *we have*

$$\left\| [(I - A)^{-1}(\tilde{x})]_T - [C_n(\tilde{x})]_T \right\|_X$$
$$\leq \frac{\|A\|^n \|[A(\tilde{x})]_T\|_X}{1 - \|A\|} \leq \frac{\|A\|_{Lip}^n \|[A(\tilde{x})]_T\|_X}{1 - \|A\|_{Lip}}.$$

Proof. First, fix $T \in [0, \infty)$. Observe that for any $\tilde{x}_1, \tilde{x}_2 \in D$, we have

$$\left\| [(I - A)(\tilde{x}_1)]_T - [(I - A)(\tilde{x}_2)]_T \right\|_X$$
$$\geq \left\| [\tilde{x}_1]_T - [\tilde{x}_2]_T \right\|_X - \left\| [A(\tilde{x}_1)]_T - [A(\tilde{x}_2)]_T \right\|_X$$
$$\geq (1 - \|A\|) \left\| [\tilde{x}_1]_T - [\tilde{x}_2]_T \right\|_X .$$

Dividing both sides by a nonzero $\left\| [\tilde{x}_1]_T - [\tilde{x}_2]_T \right\|_X$ and then taking supremum over $[0, \infty)$ with $[\tilde{x}_1]_T \neq [\tilde{x}_2]_T$, we see that $I - A$ is injective on D. Hence, in order to prove that $I - A$ is invertible and is Lipschitz on D, we have to further show that it is surjective on D and verify that $(I - A)^{-1} \in Lip(D)$.

Again, let $T \in [0, \infty)$ be fixed. We first prove that if $(I - A)^{-1}$ exists, then it is in $Lip(D)$. For any $x_1, x_2 \in D$, let $\tilde{x}_1, \tilde{x}_2 \in D$ be such that $x_1(t) = (I - A)(\tilde{x}_1)(t)$ and $x_2(t) = (I - A)(\tilde{x}_2)(t)$ for all $t \in [0, \infty)$. Then, since $\|A\| < 1$, it follows from the above inequality that if $(I - A)^{-1}$ exists then

$$\left\| [(I - A)^{-1}(x_1)]_T - [(I - A)^{-1}(x_2)]_T \right\|_X$$
$$= \left\| [\tilde{x}_1]_T - [\tilde{x}_2]_T \right\|_X$$
$$\leq (1 - \|A\|)^{-1} \left\| [(I - A)(\tilde{x}_1)]_T - [(I - A)(\tilde{x}_2)]_T \right\|_X$$
$$= (1 - \|A\|)^{-1} \left\| [x_1]_T - [x_2]_T \right\|_X ,$$

which implies that the inverse mapping of $I - A$ (if it exists) is in $Lip(D)$ with

$$\|(I - A)^{-1}\|_{Lip} \leq \|(I - A)^{-1}(x_0)\|_X + (1 - \|A\|)^{-1} .$$

Next, we prove that $I - A$ is surjective on D so that $(I - A)^{-1}$ does exist on D. For any fixed $T \in [0, \infty)$ and fixed $\tilde{x} \in D$, by definition of the operators C_n, we can show that

$$\left\| [C_{n+1}(\tilde{x})]_T - [C_n(\tilde{x})]_T \right\|_X \le \|A\|^n \left\| [A(\tilde{x})]_T \right\|_X , \quad n = 0, 1, 2, \dots .$$

Indeed, for $n = 0$, it is obviously true. Suppose that this inequality holds for $n = k - 1$. Then, since the operator A is Lipschitz, we have

$$\begin{aligned}
&\left\| [C_{k+1}(\tilde{x})]_T - [C_k(\tilde{x})]_T \right\|_X \\
&= \left\| [A \circ C_k(\tilde{x})]_T - [A \circ C_{k-1}(\tilde{x})]_T \right\|_X \\
&\le \|A\| \left\| [C_k(\tilde{x})]_T - [C_{k-1}(\tilde{x})]_T \right\|_X \\
&\le \|A\| \|A\|^{k-1} \left\| [A(\tilde{x})]_T \right\|_X ,
\end{aligned}$$

so that the inequality is true for all $n = 0, 1, 2, \dots$. Consequently, for any positive integer m,

$$\begin{aligned}
&\left\| [C_{n+m}(\tilde{x})]_T - [C_n(\tilde{x})]_T \right\|_X \\
&= \left\| \sum_{k=0}^{m-1} \left([C_{n+k+1}(\tilde{x})]_T - [C_{n+k}(\tilde{x})]_T \right) \right\|_X \\
&\le \sum_{k=0}^{m-1} \left\| [C_{n+k+1}(\tilde{x})]_T - [C_{n+k}(\tilde{x})]_T \right\|_X \\
&\le \sum_{k=0}^{m-1} \|A\|^{n+k} \left\| [A(\tilde{x})]_T \right\|_X \\
&\le \frac{\|A\|^n \left\| [A(\tilde{x})]_T \right\|_X}{1 - \|A\|} .
\end{aligned}$$

Since $\|A\| < 1$, the above inequality implies that $\{[C_n(\tilde{x})]_T\}$ is a Cauchy sequence in D, so that $\lim_{n \to \infty} [C_n(\tilde{x})]_T = \tilde{v}_T$ in norm for some $\tilde{v}_T \in D$. Let $C : \tilde{x} \mapsto \tilde{v}$ be the nonlinear operator such that $[C(\tilde{x})]_T = \tilde{v}_T$ for all $T \in [0, \infty)$. Then, C has domain D since $\tilde{x} \in D$ is arbitrary and C is independent of T. Moreover, we have $\lim_{n \to \infty} [C_n(\tilde{x})]_T = [C(\tilde{x})]_T$ for all $\tilde{x} \in D$. We will furthermore show that $C = (I - A)^{-1}$. In doing so, observe that for any $\tilde{x} \in D$, it follows from the above inequality that

$$\left\| [C(\tilde{x})]_T - [C_n(\tilde{x})]_T \right\|_X = \lim_{m \to \infty} \left\| [C_{n+m}(\tilde{x})]_T - [C_n(\tilde{x})]_T \right\|_X$$
$$\leq \frac{\|A\|^n \left\| [A(\tilde{x})]_T \right\|_X}{1 - \|A\|},$$

and note that A is Lipschitz and hence is continuous. Consequently, we have

$$[C(\tilde{x})]_T = \lim_{n \to \infty} [C_n(\tilde{x})]_T$$
$$= \lim_{n \to \infty} [(I + A \circ C_{n-1})(\tilde{x})]_T$$
$$= \tilde{x}_T + [A \circ C(\tilde{x})]_T,$$

where again the convergence is in norm. Since this holds for all $T \in [0, \infty)$, we have $C = I + A \circ C$ on D, so that $(I - A) \circ C = I$, which implies that C is a right inverse of $I - A$. Note that this also implies that $I - A$ is surjective. In summary, we have verified that $C = (I - A)^{-1} \in Lip(D)$. The rest is obvious from the above derivation.   □

Finally in this section, we remark that many important results for Lipschitz operators, such as the useful Corollary 2.5, can also be carried over to generalized Lipschitz operators. This can be easily verified by a routine procedure.

§ 5. Differentiable Mappings

Recall from Definition 2.1 that a Lipschitz operator is always continuous on its domain. However, it may not be differentiable in the sense of Fréchet defined in 1.21. As is known from elementary real analysis, differentiable mappings have more elegant properties than the continuous ones. Indeed, differentiable mappings turn out to be more useful in systems and control engineering applications. Hence, it is worth studying in some detail those differentiable Lipschitz operators. For this purpose, this section is devoted to a brief investigation of the nonlinear Lipschitz operators that are also differentiable in the sense of Fréchet (or simply, F-differentiable) on their domains.

We first need some new notations. Let X and Y be Banach function spaces over the field $\mathcal{C}$ of complex numbers. Let $D \subseteq X$ be an open and convex subset. Let $\mathcal{A} = \mathcal{A}(D, Y)$ be the family of operators $A : D \to Y$ defined by

$$\mathcal{A} = \{A : D \to Y : \quad A'(x) \text{ exists for each } x \in D$$
$$\text{and } \sup_{x \in D} \|A'(x)\|_Y < \infty \}, \qquad (2.19)$$

where $A'(x)$ is the F-derivative of A at x. Moreover, let $\mathcal{A}_c = \mathcal{A}_c(D, Y)$ be the family of continuously differentiable operators $A : D \to Y$ defined by

$$\mathcal{A}_c = \{ A \in \mathcal{A} : \quad A' \text{ is continuous on } D \}. \qquad (2.20)$$

We first have the following result, which can be easily verified.

2.23. Proposition. $\mathcal{A}$ and $\mathcal{A}_c$ are both linear spaces over the field $\mathcal{C}$ of complex numbers.

Moreover, we have the following:

2.24. Theorem. Let $Lip(D, Y)$ be the family of Lipschitz operators defined as in 2.1. Then, $A \in \mathcal{A}(D, Y)$ if and only if A is F-differentiable on D and $A \in Lip(D, Y)$. Moreover, if $A \in \mathcal{A}(D, Y)$

with the operator semi-norm defined by

$$\|A\| = \sup_{\substack{x_1, x_2 \in D \\ x_1 \neq x_2}} \frac{\|A(x_1) - A(x_2)\|_Y}{\|x_1 - x_2\|_X},$$

then we have

$$\|A\| = \sup_{x \in D} \|A'(x)\|,$$

where $A'(x)$ is the F-derivative of A at $x \in D$. Finally, $\mathcal{A}(D, Y)$ is a closed subspace of the Banach space $Lip(D, Y)$ under the Lipschitz norm $\|\cdot\|_{Lip}$. Consequently, $\mathcal{A}(D, Y)$ is a Banach space under the norm $\|\cdot\|_{Lip}$.

We remark that since $\|\cdot\|$ is only a semi-norm in general, $\mathcal{A}(D, Y)$ may not be complete under $\|\cdot\|$. Hence, the theorem is incorrect if the norm $\|\cdot\|_{Lip}$ is replaced by the semi-norm $\|\cdot\|$.

Proof. Let

$$L = \sup_{x \in D} \|A'(x)\|.$$

Then, $A \in \mathcal{A}(D, Y)$ implies $L < \infty$, so that by the F-differentiability of A, we have

$$\|A(x_1) - A(x_2)\|_Y \leq L\|x_1 - x_2\|_X$$

for all $x_1, x_2 \in D$. Hence, $A \in Lip(D, Y)$. Now, suppose that $A \in Lip(D, Y)$. Then, since A is F-differentiable by assumption, $A'(x)$ exists for each $x \in D$ and is a bounded linear operator in $\mathcal{L}(D, Y)$ by Theorem 1.25, so that

$$\|A\| = \sup_{\substack{x_1, x_2 \in D \\ x_1 \neq x_2}} \frac{\|A(x_1) - A(x_2)\|_Y}{\|x_1 - x_2\|_X}$$

$$\leq L = \sup_{x \in D} \|A'(x)\| < \infty.$$

For $\epsilon > 0$, let $x \in D$ be such that $\|A'(x)\| \geq L - \epsilon$. For such an $x \in D$ and any $x_0 \in X$ satisfying $\|x_0\|_X = 1$ and $\|A'(x)(x_0)\|_Y \geq (L - \epsilon) - \epsilon$, we have

$$
\begin{aligned}
\|A\| &\geq \varlimsup_{\lambda \to 0+} \frac{\|A(x + \lambda x_0) - A(x)\|_Y}{\|\lambda x_0\|_X} \\
&\geq \varlimsup_{\lambda \to 0+} \frac{\|A'(x)(\lambda x_0)\|_Y}{\|\lambda x_0\|_X} \\
&\quad - \frac{\|A(x + \lambda x_0) - A(x) - A'(x)(\lambda x_0)\|_Y}{\|\lambda x_0\|_X} \\
&= \frac{\|A'(x)(x_0)\|_Y}{\|x_0\|_X} \\
&\geq (L - \epsilon) - \epsilon,
\end{aligned}
$$

which implies that $L \leq \|A\|$. Hence, we have $\|A\| = L$ and $A \in \mathcal{A}(D, Y)$, as claimed.

To verify the last assertion, let $\{A_k\}$ be a sequence in $\mathcal{A}(D, Y)$ and let $A \in Lip(D, Y)$ be such that $\|A_k - A\|_{Lip} \to 0$ as $k \to \infty$. Then, by what we have just proved, we first have that

$$
\sup_{x \in D} \|A'_n(x) - A'_m(x)\| = \|A_n - A_m\| \leq \|A_n - A_m\|_{Lip} \to 0
$$

as $m, n \to \infty$. Since $A'(x)$ is a bounded linear operator in $\mathcal{L}(D, Y)$ for each fixed $x \in D$ by Theorem 1.25, this implies that for each $x \in D$, $\{A'_k(x)\}$ is a Cauchy sequence and hence converges to some operator $B(x)$ in $\mathcal{L}(D, Y)$. On the other hand, it implies that the sequence $\{A'_k(x)\}$ converges uniformly on (bounded closed subset of) D. Based on these facts, we will show that $A_k(x) \to A(x)$ uniformly on each bounded closed subset of D, A is F-differentiable on D, and $A'(x) = B(x)$ for all $x \in D$, so that $A \in \mathcal{A}(D, Y)$, or in other words, $\mathcal{A}(D, Y)$ is a closed subspace of $Lip(D, Y)$. Indeed, since $\|A_k(x) - A(x)\|_Y \leq \|A_k - A\|_{Lip}\|x\|_X \to 0$ as $k \to \infty$ for each $x \in D$, the sequence $\{A_k\}$ converges pointwise on D. Moreover, if $\{A_k\}$ converges at a point $x_1 \in D$, then since

$$
\begin{aligned}
\|A_n(x_1) &- A_m(x_1) - (A_n(x_2) - A_m(x_2))\|_Y \\
&\leq \|x_1 - x_2\|_X \sup_{x \in D} \|A'_n(x) - A'_m(x)\|,
\end{aligned}
\tag{2.21}
$$

$\{A_k\}$ converges at any point x_2 in any bounded subset $\tilde{D}$ of D. To see that A is F-differentiable on D with $A'(x) = B(x)$ for all

$x \in D$, we first observe that for any given $\epsilon > 0$, since $\{A'_k(x)\}$ is a Cauchy sequence converging to $B(x)$, there exists an integer N such that for $n, m \geq N$, we have $\|A'_n(x) - A'_m(x)\| \leq \epsilon/\delta$, where δ is the radius of a ball that covers the bounded subset $\tilde{D}$ of D (recall that D is convex), and moreover $\|B(x) - A'_n(x)\| \leq \epsilon$ for any $x \in \tilde{D}$. Consequently by letting $m \to \infty$ in the inequality (2.21), we see that for $n \geq N$ and for any $x_1, x_2 \in \tilde{D}$,

$$\|A(x_1) - A(x_2) - (A_n(x_1) - A_n(x_2))\|_Y \leq \epsilon,$$

so that

$$
\begin{aligned}
\|A(x_1) &- A(x_2) - B(x_2)(x_1 - x_2)\|_Y \\
&\leq \|A(x_1) - A(x_2) - (A_n(x_1) - A_n(x_2))\|_Y \\
&\quad + \|(A_n(x_1) - A_n(x_2)) - A'_n(x_2)(x_1 - x_2)\|_Y \\
&\quad + \|A'_n(x_2)(x_1 - x_2) - B(x_2)(x_1 - x_2)\|_Y \\
&\leq 3\epsilon,
\end{aligned}
$$

which implies that $A'(x_2)$ exists and is equal to $B(x_2)$ for all $x_2 \in \tilde{D}$. Since this proof works for any $x = x_2 \in D$ with an arbitrary bounded subset $\tilde{D}$ containing x, we have actually proved that A is F-differentiable with $A'(x) = B(x)$ for all $x \in D$. □

Clearly, Theorem 2.24 holds if we replace $\mathcal{A}(D, Y)$ by $\mathcal{A}_c(D, Y)$ therein. Hence, we may actually conclude the following:

2.25. Theorem. *Let X and Y be two Banach spaces with an open and convex subset $D \subseteq X$. Then, each of the four spaces $\mathcal{L}(D, Y), \mathcal{A}(D, Y), \mathcal{A}_c(D, Y)$, and $Lip(D, Y)$ are Banach spaces under the Lipschitz norm $\|\cdot\|_{Lip}$, and moreover,*

$$\mathcal{L}(D, Y) \subset \mathcal{A}_c(D, Y) \subset \mathcal{A}(D, Y) \subset Lip(D, Y).$$

The following result is immediate.

2.26. Corollary. *Let X and Y be two Banach spaces. Then, each of the four spaces $\mathcal{L}(X, Y), \mathcal{A}(X, Y), \mathcal{A}_c(X, Y)$, and $Lip(X, Y)$ are Banach spaces under the Lipschitz norm $\|\cdot\|_{Lip}$, and*

$$\mathcal{L}(X, Y) \subset \mathcal{A}_c(X, Y) \subset \mathcal{A}(X, Y) \subset Lip(X, Y).$$

In particular, if $X = Y$, then all the spaces $\mathcal{L}(X), \mathcal{A}_c(X), \mathcal{A}(X)$, and $Lip(X)$ are Banach spaces and are near-rings with unity I.

Finally, we close this section by the following important result.

2.27. Theorem. *Let X be a Banach space and $D \subseteq X$ be an open and convex subset. Suppose that A is an invertible element in $Lip(X)$ and that for a fixed $x \in D$, the F-derivative $A'(x)$ exists. Then, $A'(x)$ is an invertible element in $\mathcal{L}(X)$.*

Proof. Let $C(x_0) = A(x + x_0) - A(x)$ for all $x_0 \in X$. Then, it is easily seen that $C, C^{-1} \in Lip(X), C(0) = 0$, and (the bounded linear operator) $C'(0) = A'(x)$. Consequently, we have

$$\lim_{h \to 0} \frac{\|C(h)\|_X}{\|h\|_X} = \lim_{h \to 0} \frac{\|A'(x)(h)\|_X}{\|h\|_X},$$

where it should be noted that $A'(x)$ on the right-hand side is a bounded linear operator, so that the limit is only a formal operation. Now, let $x_0 \in X$ with $\|x_0\|_X = 1$. Since $A'(x)$ is linear, it follows from the above formal equality that

$$\|(A'(x))'(x_0)\|_X = \lim_{\lambda \to 0+} \frac{\|A'(x)(\lambda x_0)\|_X}{\|\lambda x_0\|_X} = \lim_{\lambda \to 0+} \frac{\|C(\lambda x_0)\|_X}{\|\lambda x_0\|_X}.$$

Then, observe that since A is invertible in $Lip(X)$, we have $0 < \|C^{-1}\|_{Lip} < \infty$, so that

$$\|\tilde{x}\|_X = \|C^{-1}(C(\tilde{x}))\|_X \leq \|C^{-1}\| \, \|C(\tilde{x})\|_X,$$

or

$$\frac{\|C(\tilde{x})\|_X}{\|\tilde{x}\|_X} \geq \|C^{-1}\|^{-1}$$

for all $\tilde{x} \in X$. It follows that

$$\|(A'(x))'(x_0)\|_X \geq \|C^{-1}\|^{-1}$$

for all $x_0 \in X$ with $\|x_0\|_X = 1$. Consequently,

$$\|A'(x)(\tilde{x})\|_X \geq \|C^{-1}\|^{-1} \|\tilde{x}\|_X$$

for all $\tilde{x} \in X$. This implies that $A'(x)$ is injective. Moreover, it can be shown that the range of $A'(x)$ is closed. As a matter of fact, if

$\{x_k\}$ and $\{y_k\}$ are two sequences in X such that $y_k = A'(x)(x_k)$ and $\|y_k - y\| \to 0$ as $k \to \infty$, then since

$$\|x_n - x_m\|_X \le \|C^{-1}\| \|A'(x)(x_n) - A'(x)(x_m)\|_X$$
$$\le \|C^{-1}\| \|y_n - y_m\|_X \to 0$$

as $n, m \to \infty$, and since X is complete, we have $x_k \to x^*$ for some $x^* \in X$. Hence,

$$A'(x)(x^*) = \lim_{k \to \infty} A'(x)(x_k) = \lim_{k \to \infty} y_k = y\,,$$

which implies that the range of $A'(x)$ is closed. Furthermore, it can be shown that the range of $A'(x)$ is dense in X. In doing so, let $\epsilon > 0$ and let $x_0 \in X$ with $\|x_0\|_X = 1$. Choose a $\delta > 0$ such that $\|C(h) - A'(h)\|_X \le \epsilon \|h\|_X$ for all $\|h\|_X \le \delta$. Then, for any $\tilde{x} \in X$ satisfying $C(\tilde{x}) = \delta \|C^{-1}\|^{-1} x_0$ (note that C is continuous with $C(0) = 0$ and so $\tilde{x}$ always exists), we have

$$\|\tilde{x}\|_X = \|C^{-1}(\delta \|C^{-1}\|^{-1} x_0)\|_X$$
$$\le \|C^{-1}\| \delta \|C^{-1}\|^{-1} \|x_0\|_X$$
$$= \delta\,,$$

so that by the choice of δ we obtain

$$\|C(\tilde{x}) - A'(x)(\tilde{x})\|_X \le \epsilon \|\tilde{x}\|_X \le \epsilon \delta\,.$$

Consequently,

$$\|x_0 - A'(x)(\delta^{-1}\|C^{-1}\|\tilde{x})\|_X$$
$$= \delta^{-1}\|C^{-1}\| \|C(\tilde{x}) - A'(x)(\tilde{x})\|_X$$
$$\le \|C^{-1}\|\epsilon\,.$$

It then follows that for each nonzero $y \in X$, there exists an $x_y \in X$ such that

$$\big\| \|y\|_X^{-1} y - A'(x)x_y \big\|_X \le \|C^{-1}\|\epsilon$$

or

$$\|y - A'(x)(\|y\|_X x_y)\|_X \le \|y\|_X \|C^{-1}\|\epsilon\,.$$

This implies that the range of $A'(x), \mathcal{R}(A'(x))$ is dense in X. Since we have proved that $\mathcal{R}(A'(x))$ is closed in X, we have $\mathcal{R}(A'(x)) = X$. Hence, $A'(x)$ is both injective and surjective, and is hence invertible. Finally, since

$$\|A'(x)(\tilde{x})\|_X \geq \|C^{-1}\|^{-1}\|\tilde{x}\|_X$$

for all $\tilde{x} \in X$, we have $(A'(x))^{-1} \in \mathcal{L}(X)$ with $\|(A'(x))^{-1}\| \leq \|C^{-1}\|$. $\qquad \square$

Exercises

2.1. Give some examples of nonlinear Lipschitz operators and of invertible nonlinear Lipschitz operators.

2.2. Let X and Y be normed linear spaces and $A : \mathcal{D}(A) \to Y$ be a nonlinear operator, where $\mathcal{D}(A)$ is the domain of A contained in X. Show that

$$\|A\| := \|A(0)\|_Y + \sup_{\substack{x \in \mathcal{D}(A) \\ x \neq 0}} \frac{\|A(x) - A(0)\|_Y}{\|x\|_X}$$

defines a norm for A. Compare this norm with the Lipschitz norm defined in Theorem 2.3.

2.3. Give some examples to show that for a non-Lipschitz operator $A : X \to X$, where X is a normed linear space, there can be an integer $n > 1$ such that A^n is Lipschitz on X.

2.4. Generalize the result of Examples 2.17 and 2.18 to the n-dimensional space $L_p^n[0, T]$, with $1 \leq p \leq \infty$, as remarked at the end of Section 3.

2.5. Let X^e, Y^e, Z^e be extended linear spaces and D be a subset of X^e. Denote by $Lip(D, Y^e)$ the family of generalized Lipschitz operators A mapping from D to Y^e with $\mathcal{D}(A) = D$ and $\mathcal{R}(A)$ being a subspace of Y^e, and let $C : \mathcal{R}(A) \to Z^e$ be a bounded (in any operator norm) nonlinear operator with $\mathcal{D}(C) = \mathcal{R}(A)$ and $\mathcal{R}(C) \subseteq Z^e$. Moreover, set

$$S = \{ A \in Lip(D, Y^e) : \quad C \circ A \in Lip(D, Z^e) \} .$$

Show that S is a Banach space.

References

[1] G. Chen and R.J.P. de Figueiredo, "On construction of co-prime factorizations of nonlinear feedback control systems," *Circ. Sys. Sign. Proc.*, **11** (1992), 285–307.

[2] J. Dieudonné, *Foundations of Modern Analysis*, Academic Press, New York, 1960.

[3] V. Dolezal, *Monotone Operators and Applications in Control and Network Theory*, Elsevier, New York, 1979.

[4] R.H. Martin, *Nonlinear Operators and Differential Equations in Banach Spaces*, Wiley, New York, 1976.

[5] E. Zeidler, *Nonlinear Functional Analysis and its Applications*, Springer-Verlag, New York, 1986.

Chapter 3

Nonlinear Feedback Systems

Some necessary fundamental developments in mathematics have been presented in the previous two chapters. Supported by these foundations, the current chapter will be devoted to the development of a general framework for the analysis of nonlinear feedback control systems. The framework to be introduced is established in a general extended linear space setting with the control systems described by nonlinear operators. Several fundamental issues for nonlinear feedback systems such as causality and stability, as well as well-posedness in the sense of the uniqueness of internal control signals, will be studied. It will be shown that generalized nonlinear Lipschitz operators are always causal under the extended linear space framework. The so-called small gain theorem, together with its strengthened version, will then be discussed. Next, important notions pertaining to stability will be studied in detail, and moreover, those feedback compensators that stabilize the overall closed-loop control system will be characterized in the Lipschitz-norm setting, which will ensure as well the causality and well-posedness of the system. Finally, a relationship between the bounded-input/bounded-output (BIBO) stability and the Lyapunov stability for a nonlinear control system described by the standard state-variable representation will be discussed.

§ 1. Causality of Feedback Systems

A general feedback control system in engineering applications is described by the closed-loop configuration shown in Figure 3.1, where S_1, S_2 denote the two nonlinear subsystems under investigation, u_1, u_2 two control inputs, y_1, y_2 the corresponding system outputs, and e_1, e_2 the error signals. A more precise description of this closed-loop system will be given later.

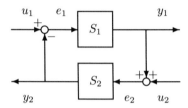

Figure 3.1.   A nonlinear feedback system.

This feedback system may also be viewed as the one shown in Figure 3.2, where P represents a plant and C a compensator, u_2 may be considered to be a disturbance input, and e_2 the actual system output. Many different ways of looking at the configuration shown in Figure 3.1 are possible, depending on the structure of the real system and the purpose of the investigation.

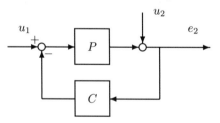

Figure 3.2.   A nonlinear feedback system.

For a continuous-time feedback system such as the one shown in Figure 3.1, all the inputs, outputs, and internal control signals are functions of time, $t \in [0, \infty)$, taking values in certain vector-valued Banach spaces. Let us denote by X_1 a p-dimensional Banach space of measurable functions defined on the time domain $[0, \infty)$

to which signals u_1, e_1, and y_2 belong, and by X_2 a q-dimensional Banach space similarly defined to which u_2, e_2, and y_1 belong. In this case, the closed-loop configuration is usually called a *multi-input/multi-output* (MIMO) feedback system. For simplicity of notation, however, we will mainly consider a scalar-valued setting of the system, usually called a *single-input/single-output* (SISO) feedback system, in the following two sections. It will be seen that all the results and derivations given below can be easily converted to MIMO systems.

In order to develop some optimal feedback control systems design strategies for a rather general situation later, we will consider the generalized Lipschitz operators introduced and discussed in Chapter 2, Section 4, under an extended linear space framework. Recall that one reason for considering extended linear spaces is that all the control signals in real life are time-limited, so that an extended linear space is particularly suitable. In addition, if we take the so-called *causality* into account, which is a basic requirement for realizing a physical system, then the extended linear space framework is advantageous. To see this, we first introduce the concept of causality.

3.1. Definition. Let X^e be the extended linear space associated with a given Banach space X, and let $S : X^e \to X^e$ be a nonlinear mapping describing a nonlinear control system. Then, S is said to be *causal* if and only if

$$P_T S P_T = P_T S$$

for all $T \in [0, \infty)$, where P_T is the projection operator with respect to T defined as before by

$$P_T(x(t)) = \begin{cases} x(t) & 0 \le t \le T \\ 0 & T < t < \infty \end{cases}$$

for all $x(t) \in X^e$.

The physical meaning behind this definition of causality may be understood as follows. If the system outputs depend only on the present and past values of the corresponding system inputs,

---

then we have $SP_T(u) = S(u)$ for all the input signals u in the domain of S, so that $P_T SP_T = P_T S$. Conversely, if $P_T SP_T = P_T S$ for all $T \in [0, \infty)$, then we have $P_T S(I - P_T)(u) = 0$ for all u in the domain of S, which implies that any future value of a system input, $(I - P_T)(u)$, does not affect the present and past values of the corresponding system output given by $P_T S(\cdot)$, or in other words, system outputs depend only on the present and past values of the corresponding system inputs.

3.2. Example. Let M be the family of complex-valued measurable functions defined on $[0, \infty)$ and $X = L_\infty(0, \infty)$ be defined as

$$X = \left\{ f \in M : \ ess \sup_{0 \le t < \infty} |f(t)| < \infty \right\}.$$

Let

$$X^e = \left\{ f \in M : \ ess \sup_{0 \le t \le T} |f(t)| < \infty \ \text{for all} \ T < \infty \right\}.$$

Then, the integral operator $S : X^e \to X^e$ defined below is causal:

$$S(x)(t) = \int_0^t K(t, \tau, x(\tau))d\tau,$$

provided that K is such that $K(\cdot, \cdot, x) \to K(t, \tau, x)$ is integrable on $[0, \infty) \times [0, \infty)$ and $K(t, \tau, \cdot) \to K(t, \tau, x)$ is continuous on X^e.

For any function f defined on $[0, \infty)$, let

$$[f]_T = P_T(f(t))$$

be the truncation of f with respect to $T, T \in [0, \infty)$. We first have the following result.

3.3. Theorem. *A nonlinear mapping $S : X^e \to X^e$ is causal if and only if for any $x, y \in X^e$ and any $T \in [0, \infty), x_T = y_T$ implies $[S(x)]_T = [S(y)]_T$.*

Proof. Suppose that S is causal. Then by definition we have $P_T S P_T = P_T S$, so that if $x_T = y_T$ then

$$[S(x)]_T = P_T S(x) = P_T S P_T(x) = P_T S(x_T) = P_T S(y_T)$$
$$= P_T S P_T(y) = P_T S(y) = [S(y)]_T.$$

Conversely, suppose that $x_T = y_T$ implies $[S(x)]_T = [S(y)]_T$ for all $x, y \in X^e$ and all $T \in [0, \infty)$. Fix a $T \in [0, \infty)$. For any $x \in X^e$, let $y = x_T$. Then we have $y_T = x_T$, so that $[S(x)]_T = [S(y)]_T$. Consequently, we have

$$P_T S P_T(x) = P_T S(x_T) = P_T S(y)$$
$$= [S(y)]_T = [S(x)]_T = P_T S(x).$$

Since $x \in X^e$ and $T \in [0, \infty)$ are arbitrary, it follows that $P_T S P_T = P_T S$ for all $T \in [0, \infty)$, which implies that S is causal. $\square$

3.4. Corollary. *Let $S : X^e \to X^e$ be a nonlinear generalized Lipschitz operator as defined in 2.19. Then S is causal.*

To prove this result, it is sufficient to note that

$$\|[S(x)]_T - [S(y)]_T\|_X \le \|S\|_{Lip} \|x_T - y_T\|_X \qquad (3.1)$$

for all $x, y \in X^e$ and all $T \in [0, \infty)$. Hence, $x_T = y_T$ implies $[S(x)]_T = [S(y)]_T$ for all $x, y \in X^e$ and all $T \in [0, \infty)$.

Note that a nonlinear operator may produce nonunique outputs from an input, particularly for a set-valued mapping. For the well-posedness of a control system, we usually require all internal signals be unique. It is clear from (3.1) that a nonlinear generalized Lipschitz operator always guarantees this uniqueness requirement. More precisely, we have the following:

3.5. Corollary. *A nonlinear generalized Lipschitz operator produces a unique output from an input in the sense that if the input x and output y are related by a generalized Lipschitz operator S such that $y = S(x)$, then $x_T = \tilde{x}_T$ implies $y_T = \tilde{y}_T$ for all $T \in [0, \infty)$.*

§ 2. The Small Gain Theorem

In this section, we will as usual let X^e be the extended linear space associated with a given Banach space X of complex-valued measurable functions defined on the time domain $[0, \infty)$.

Recall that one reason for using the extended linear space framework is that if the norm $\|f_T\|_X$ is monotonically increasing as $T \to \infty$ for all $f \in X^e$ and satisfies $\|f_T\|_X \to \|f\|_X$ as $T \to \infty$ for all $f \in X$, then many useful techniques and results may be carried over from X to X^e. Hence, here and throughout, we will impose these two mild conditions, where it should be noted that almost all common norms satisfy these two conditions.

Now, let us consider the nonlinear feedback system shown in Figure 3.1. Assume that $u_1, u_2 \in X^e$ and that $S_1 : X^e \to X^e$ and $S_2 : X^e \to X^e$. It is clear from the connection of the closed-loop configuration shown in Figure 3.1 that

$$\begin{cases} e_1 = u_1 - S_2(e_2) \\ e_2 = u_2 + S_1(e_1), \end{cases} \tag{3.2}$$

where $e_1, e_2 \in X^e$. Equivalently, we have

$$\begin{cases} u_1 = e_1 + S_2(e_2) \\ u_2 = e_2 - S_1(e_1). \end{cases} \tag{3.3}$$

For this system, independently of whether S_1 and S_2 are causal or not, we have the following so-called *small gain theorem*, which gives sufficient conditions under which a "bounded input" yields a "bounded output," where the boundedness is in the norm of X.

3.6. Theorem. *Consider the nonlinear feedback system shown in Figure 3.1, which is described by the relationships in (3.3). Suppose that there exist constants L_1, L_2, M_1, M_2, with $L_1, L_2 \geq 0$ and $L_1 L_2 < 1$, such that*

$$\begin{cases} \left\| [S_1(e_1)]_T \right\|_X \leq M_1 + L_1 \left\| [e_1]_T \right\|_X \\ \left\| [S_2(e_2)]_T \right\|_X \leq M_2 + L_2 \left\| [e_2]_T \right\|_X \end{cases} \tag{3.4}$$

for all $T \in [0, \infty)$. Then, we have

$$
\begin{cases}
\|[e_1]_T\|_X \leq (1 - L_1 L_2)^{-1} \big(\|[u_1]_T\|_X \\
\qquad\qquad + L_2 \|[u_2]_T\|_X + M_2 + L_2 M_1 \big) \\[2mm]
\|[e_2]_T\|_X \leq (1 - L_1 L_2)^{-1} \big(\|[u_2]_T\|_X \\
\qquad\qquad + L_1 \|[u_1]_T\|_X + M_1 + L_1 M_2 \big)
\end{cases}
\tag{3.5}
$$

for all $T \in [0, \infty)$. If, in addition, $\|u_1\|_X < \infty$ and $\|u_2\|_X < \infty$, then the inequalities in (3.5) hold with all the subscripts T therein being dropped.

We remark that constants L_1, L_2, M_1, and M_2 are usually not unique. However, if S_1 is a generalized Lipschitz operator defined on X^e, then we may precisely choose $M_1 = \|S_1(0)\|_X$ and

$$
L_1 = \sup_{T \in [0, \infty)} \; \sup_{\substack{e_{11}, e_{12} \in X^e \\ [e_{11}]_T \neq [e_{12}]_T}} \frac{\|[S_1([e_{11}]_T)]_T - [S_1([e_{12}]_T)]_T\|_X}{\|[e_{11}]_T - [e_{12}]_T\|_X}.
$$

The same remark applies to the operator S_2.

We also note that the condition $L_1, L_2 \geq 0$ is not necessary in establishing the inequalities in (3.5), as will be seen below in the proof. However, since L_1 and L_2 play the role of the gain of an operator, this condition is natural.

Proof. It follows from (3.2) that

$$
[e_1]_T = [u_1]_T - [S_2(e_2)]_T \,,
$$

so that

$$
\begin{aligned}
\|[e_1]_T\|_X &\leq \|[u_1]_T\|_X + \|[S_2(e_2)]_T\|_X \\
&\leq \|[u_1]_T\|_X + M_2 + L_2 \|[e_2]_T\|_X
\end{aligned}
$$

for all $T \in [0, \infty)$. Similarly, we have

$$
\|[e_2]_T\|_X \leq \|[u_2]_T\|_X + M_1 + L_1 \|[e_1]_T\|_X
$$

for all $T \in [0, \infty)$. Combining these two inequalities, we obtain

$$\|[e_1]_T\|_X \leq L_1 L_2 \|[e_1]_T\|_X + \|[u_1]_T\|_X$$
$$+ L_2 \|[u_2]_T\|_X + M_2 + L_2 M_1,$$

or, using the fact that $L_1 L_2 < 1$,

$$\|[e_1]_T\|_X \leq (1 - L_1 L_2)^{-1} \big(\|[u_1]_T\|_X$$
$$+ L_2 \|[u_2]_T\|_X + M_2 + L_2 M_1 \big).$$

The rest of the theorem follows immediately. $\square$

It is clear that the small gain theorem is applicable to both continuous-time and discrete-time systems and both SISO and MIMO systems. Hence, although its statement and proof are quite simple, it is very useful.

Under rather strong conditions, we have the following strengthened form of the small gain theorem, which guarantees certain existence and uniqueness, as well as boundedness and continuity.

3.7. Theorem. *Consider the nonlinear feedback system shown in Figure 3.1, which is described by the relationships in (3.3). Fix a $T \in [0, \infty)$. Suppose that the linear subspace of X^e defined by*

$$X_T^e := \{ f_T : \quad f \in X^e \}$$

is a Banach space and assume that both S_1 and S_2 are generalized Lipschitz operators with

$$\begin{cases} \|[S_1(x_1)]_T - [S_1(x_2)]_T\|_X \leq L_1 \|[x_1]_T - [x_2]_T\|_X \\ \|[S_2(x_1)]_T - [S_2(x_2)]_T\|_X \leq L_2 \|[x_1]_T - [x_2]_T\|_X \end{cases} \quad (3.6)$$

for all $x_1, x_2 \in X^e$, where $L_1 L_2 < 1$. Then, for any given inputs $u_1, u_2 \in X^e$, the system outputs y_1, y_2 and error signals e_1, e_2 are uniquely determined by u_1, u_2 via

$$\begin{cases} [e_1]_T = [u_1]_T - [S_2(u_2 + S_1(e_1))]_T \\ [e_2]_T = [u_2]_T + [S_1(u_1 - S_2(e_2))]_T \end{cases} \quad (3.7)$$

and

$$\begin{cases} [y_1]_T = [S_1(e_1)]_T \\ [y_2]_T = [S_2(e_2)]_T \,. \end{cases} \qquad (3.8)$$

Moreover, the two system mappings $f : (u_1, u_2) \to (e_1, e_2)$ *and* $g : (u_1, u_2) \to (y_1, y_2)$ *are uniformly continuous both on* $X_T^e \times X_T^e$ *and on* $X \times X$.

Proof. The relationships (3.7) and (3.8) follow directly from (3.2) and Figure 3.1, respectively. Moreover, by Corollary 3.4, S_1 and S_2 are both causal, so that by Definition 3.1 we can rewrite (3.7) as

$$\begin{cases} [e_1]_T = [u_1]_T - [S_2([u_2]_T + [S_1([e_1]_T)]_T)]_T \\ \quad := [u_1]_T - \phi_1([e_1]_T) \\ [e_2]_T = [u_2]_T + [S_1([u_1]_T - [S_2([e_2]_T)]_T)]_T \\ \quad := [u_2]_T - \phi_2([e_2]_T) \,, \end{cases}$$

where $u_1, u_2 \in X^e$ are given inputs. Now, it can be verified that ϕ_1 and ϕ_2 are both contraction mappings on the Banach space X_T^e. Indeed, for any $[u_2]_T, [e_{11}]_T, [e_{12}]_T \in X_T^e$, it follows from (3.6) that

$$\begin{aligned} &\|\phi_1([e_{11}]_T) - \phi_1([e_{12}]_T)\|_X \\ &= \|[S_2([u_2]_T + [S_1([e_{11}]_T)]_T)]_T \\ &\quad - [S_2([u_2]_T + [S_1([e_{12}]_T)]_T)]_T\|_X \\ &\leq L_2\|[S_1([e_{11}]_T)]_T - [S_1([e_{12}]_T)]_T\|_X \\ &\leq L_2 L_1\|[e_{11}]_T - [e_{12}]_T\|_X \,, \end{aligned}$$

where $L_1 L_2 < 1$ by assumption. Hence, ϕ_1 is a contraction mapping on a Banach space, so that by Theorem 2.15, $[e_1]_T$ is uniquely determined by u_1 and u_2 via the first equation in (3.7). The same holds for $[e_2]_T$.

To see that the mapping $f : (u_1, u_2) \to (e_1, e_2)$ is uniformly continuous on both $X_T^e \times X_T^e$ and $X \times X$, we rewrite the relationship $e_1 = u_1 - S_2(e_2)$ by the causality of S_2 as

$$[e_1]_T = [u_1]_T - [S_2([e_2]_T)]_T \,.$$

Then, for any $(u_1, u_2), (\tilde{u}_1, \tilde{u}_2) \in X^e \times X^e$, there exist unique solutions $(e_1, e_2), (\tilde{e}_1, \tilde{e}_2) \in X^e \times X^e$, such that

$$\left\|[e_1]_T - [\tilde{e}_1]_T\right\|_X \leq \left\|[u_1]_T - [\tilde{u}_1]_T\right\|_X + L_2\left\|[e_2]_T - [\tilde{e}_2]_T\right\|_X$$

and

$$\left\|[e_2]_T - [\tilde{e}_2]_T\right\|_X \leq \left\|[u_2]_T - [\tilde{u}_2]_T\right\|_X + L_1\left\|[e_1]_T - [\tilde{e}_1]_T\right\|_X .$$

Consequently, by combining these two inequalities we obtain

$$\left\|[e_1]_T - [\tilde{e}_1]_T\right\|_X \leq (1 - L_1 L_2)^{-1}\left(\left\|[u_1]_T - [\tilde{u}_1]_T\right\|_X \right.$$
$$\left. + L_2\left\|[u_2]_T - [\tilde{u}_2]_T\right\|_X\right)$$

and

$$\left\|[e_2]_T - [\tilde{e}_2]_T\right\|_X \leq (1 - L_1 L_2)^{-1}\left(\left\|[u_2]_T - [\tilde{u}_2]_T\right\|_X \right.$$
$$\left. + L_1\left\|[u_1]_T - [\tilde{u}_1]_T\right\|_X\right).$$

This implies that the mapping $f : (u_1, u_2) \to (e_1, e_2)$ is uniformly continuous on $X_T^e \times X_T^e$ as well as on $X \times X$. The same holds for the mapping $g : (u_1, u_2) \to (y_1, y_2)$. $\qquad\square$

It will be seen later that the small gain theorem, 3.6, and its strengthened form, 3.7, are very useful in the study of stabilization of nonlinear feedback control systems.

§ 3. Stabilities of Control Systems

Stability is one of the most important properties that a control system should possess, and is hence a major subject and a fundamental entity in the investigation of systems control. There are different types of stability for a control system. Within the scope of the present monograph, we will be mainly concerned with the so-called input-output stability and its strengthened notion of finite-gain input-output stability, where the gain of a system is referred to the operator norm of the system operator that we use.

To introduce the concept of input-output stability, let us consider a nonlinear mapping S that describes a nonlinear control system, denoted also by S, as shown in Figure 3.3, where $U \subseteq X^e$ and $V \subseteq Y^e$ are linear subspaces of the two extended linear spaces X^e and Y^e, respectively, used as part of the framework. In systems engineering terminology, U and V are called respectively the *input space* and *output space*.

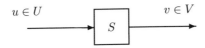

Figure 3.3.   A nonlinear system.

Note that the input and output spaces U and V may be chosen to be any normed linear spaces in general, independent of the Banach spaces X and Y used to define X^e and Y^e. But in the study of stability, causality, and many other subjects below, we will restrict $U \subseteq X^e$ and $V \subseteq Y^e$ for convenience.

A more important remark is in order. As mentioned before, by a nonlinear operator we always mean an operator that is not necessarily linear. Hence, any meaningful result for nonlinear operators should also be valid for linear operators. This is particularly important when we consider a system as an operator mapping from (a subset of) an input space to an output space. Specifically, when we consider the domain of a nonlinear operator that describes a control system, we prefer to use a linear subspace, so that when the system becomes linear, it remains to be well defined on the same domain. Of course, after defining the system domain, the admissible system inputs can be restricted to a subset in its domain. In this case, an input that is not in the admissible set is not allowed to be used in a design or in a control process, but it does not cause any trouble in view of the well-definedness and well-posedness of the system. We will accept this viewpoint in the following.

Therefore, let X^e and Y^e be two prescribed extended linear spaces associated respectively with two Banach spaces of complex-valued measurable functions on the time domain $[0, \infty)$. Let $U \subseteq X^e$ and $V \subseteq Y^e$ be the input and output spaces, respectively, of a nonlinear control system described by a nonlinear operator $S : U \to V$, where the domain of S is U. We first have the following:

3.8. Definitions. The nonlinear system S is said to be *input-output stable from U to V* if and only if it maps all inputs from U into the output space V. In notation, if and only if $S(U) \subseteq V$. Otherwise, if S maps some input from U to the set $Y^e \backslash V$ (if not empty), then the system S is said to be *unstable*.

Note that a bounded operator may map some element of U to a point that is outside the output space V. Hence, boundedness is in general not sufficient for the input-output stability. In other words, both the domain and range of the operator have to be taken into account in the investigation of the input-output stability. Besides, the norms that we used in U and V (or X and Y) are also important, because the same system S may be input-output stable under certain norms but not under some others for the same set of control inputs.

Taking the boundedness of the system operator into account, we have the following:

3.9. Definition. Let $S : U \to V$ be a nonlinear operator that describes an input-output stable control system with the input and output spaces U and V, respectively. If, furthermore, an operator norm $\|S\|$ for $S : U \to V$ is well defined and finite, $\|S\| < \infty$, then the system is said to be *finite-gain input-output stable*.

Here, it should be noted that the "gain" of the system operator depends on the operator norm used. Hence, it depends not only on how the operator norm is defined, but also on what the norms of the input and output spaces are, as mentioned above.

Now let us consider the nonlinear feedback system shown in Figure 3.1, or more precisely in Figure 3.4, where all notations are as defined above. For this closed-loop configuration consisting of two subsystems, we have the following definitions.

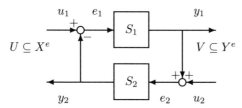

Figure 3.4. A nonlinear feedback system.

3.10. Definitions. The nonlinear feedback system shown in Figure 3.4, denoted by (S_1, S_2), is said to be *jointly input-output stable from $U \times V$ to $V \times U$* if and only if for any input pair $(u_1, u_2) \in U \times V$, the corresponding system output pair $(y_1, y_2) \in V \times U$, namely:

$$(u_1, u_2) \in U \times V \implies (y_1, y_2) \in V \times U,$$

or notationally, if and only if both $S_1(U) \subseteq V$ and $S_2(V) \subseteq U$. The system (S_1, S_2) is said to be *jointly finite-gain input-output stable* if moreover the operator norms $\|S_1\| < \infty$ and $\|S_2\| < \infty$.

Observe from Figure 3.4 that the joint input-output stability is equivalent to that of the corresponding error signal pair $(e_1, e_2) \in U \times V$ whenever the input pair $(u_1, u_2) \in (U, V)$ since both U and V are linear spaces.

Note also that when two individually input-output stable subsystems S_1 and S_2 are connected together in the feedback closed loop as shown in Figure 3.4, the resulting configuration may not be jointly input-output stable. Hence, the above conditions, say $S_1(U) \subseteq V$ and $S_2(V) \subseteq U$, are sufficient for the joint input-output stability of the overall feedback system only if these conditions can be satisfied "on-line," in the sense that they hold true in the connected closed-loop configuration.

Now, combining Theorem 3.7 and Corollary 3.4, we have the following result, which can be easily verified:

3.11. Theorem. *Consider the nonlinear feedback system shown in Figure 3.4. Suppose that $U = V = X^e$ where X^e is an extended linear space satisfying the following condition: For each fixed $T \in [0, \infty)$, the family*

$$X_T^e = \{ f_T : \quad f \in X^e \}$$

is a Banach space. Assume also that S_1 and S_2 are two generalized Lipschitz operators such that

$$\begin{cases} \left\| [S_1(e_{11})]_T - [S_1(e_{12})]_T \right\|_X \leq L_1 \left\| [e_{11}]_T - [e_{12}]_T \right\|_X \\ \left\| [S_2(e_{21})]_T - [S_2(e_{22})]_T \right\|_X \leq L_2 \left\| [e_{21}]_T - [e_{22}]_T \right\|_X \end{cases}$$

for all $e_{11}, e_{12}, e_{21}, e_{22} \in X^e$, where $L_1 L_2 < 1$. Then, the overall closed-loop system is causal, jointly finite-gain input-output stable from $U \times V$ to $V \times U$, and well posed in the sense that for any input pair $(u_1, u_2) \in X^e$, the corresponding system output pair (y_1, y_2) and error signal pair (e_1, e_2) are uniquely determined by (u_1, u_2) via

$$\begin{cases} [e_1]_T = [u_1]_T - [S_2(u_2 + S_2(e_1))]_T \\ [e_2]_T = [u_2]_T + [S_2(u_1 - S_2(e_2))]_T \end{cases} \tag{3.9}$$

and

$$\begin{cases} [y_1]_T = [S_1(e_1)]_T \\ [y_2]_T = [S_2(e_2)]_T . \end{cases} \tag{3.10}$$

Note that in a case where the admissible control inputs and the allowable system outputs are restricted only in (bounded) proper subsets $U_i \times V_i \subset U \times V$ and $U_o \times V_o \subset U \times V$, the conditions stated in the above theorem are not sufficient to ensure the jointly input-output stability of the feedback system from $U_i \times V_i$ to $V_o \times U_o$. In this case, we have to require that $S_1(U_i) \subset V_o$ and $S_2(V_i) \subset U_o$, directly referring to the definition. Hence, we have the following:

3.12. Corollary. *Consider the nonlinear feedback system shown in Figure 3.4. Let $U_i, V_o \subset U$ and $U_o, V_i \subset V$ be (bounded) proper subsets of the two extended linear spaces U and V. Assume that all conditions stated in Theorem 3.11 are satisfied, and in addition, $S_1(U_i) \subset V_o$ and $S_2(V_i) \subset U_o$. Then, the system is causal, jointly* *finite-gain input-output stable from $U_i \times V_i$ to $V_o \times U_o$, and well posed in the sense that the corresponding system output pair (y_1, y_2) and error signal pair (e_1, e_2) are uniquely determined via (3.9) and (3.10), respectively, by the input pair $(u_1, u_2) \in U_i \times V_i$.*

3.13. Example. Consider the nonlinear feedback system (S_1, S_2) shown in Figure 3.4. Let

$$U = L_p^e([0, \infty), R^n) \quad \text{and} \quad V = L_{p'}^e([0, \infty), R^m),$$

where L_p and $L_{p'}$ are standard Banach spaces of real vector-valued measurable functions defined on the time domain $[0, \infty)$, with $1 \leq p, p' \leq \infty$, and $1 \leq n, m < \infty$. Also, let $U_i = V_o = B_p$, the unit ball in $L_p^e([0, \infty), R^n)$, and $U_o = V_i = B_{p'}$, the unit ball in $L_{p'}^e([0, \infty), R^m)$. Suppose that $S_1 : L_p^e \to L_{p'}^e$ and $S_2 : L_{p'}^e \to L_p^e$ are two nonlinear Volterra integral operators defined respectively by

$$S_1(e)(t) = \int_0^t K_1(t, s, e(s)) ds, \qquad t \in [0, \infty)$$

and

$$S_2(e)(t) = \int_0^t K_2(t, s, e(s)) ds, \qquad t \in [0, \infty),$$

where the two nonlinear integrable functions K_1 and K_2 satisfy

$$\|K_1(t, s, e_{11}) - K_1(t, s, e_{12})\| \leq \alpha_1 \|e_{11} - e_{12}\|$$

and

$$\|K_2(t, s, e_{21}) - K_2(t, s, e_{22})\| \leq \alpha_2 \|e_{21} - e_{22}\|$$

for some $\alpha_1, \alpha_2 \geq 0$, all $t, s \in [0, \infty)$, all $e_{11}, e_{12} \in R^n$, and all $e_{21}, e_{22} \in R^m$, and $\| \cdot \|$ denotes the Euclidean norm of the vector-valued functions. Then, under the conditions that $\alpha_1 \alpha_2 < 1$,

$$\left\| \int_0^t K_1(t, s, e(s)) ds \right\|_{L_{p'}} \leq 1 \qquad \text{for all } e \in B_p,$$

and

$$\left\| \int_0^t K_2(t, s, e(s)) ds \right\|_{L_p} \leq 1 \qquad \text{for all } e \in B_{p'},$$

the feedback system (S_1, S_2) connected as shown in Figure 3.4 is jointly finite-gain input-output stable. This can be easily verified by reviewing Example 2.18 and checking the conditions stated in Corollary 3.12.

§ 4. Stabilizing Nonlinear Feedbacks

Consider, again, the general nonlinear feedback system shown in Figure 3.4 above. In engineering applications, such as in feedback compensator design, we are usually given a subsystem, S_1 say, and asked to design another, S_2, to satisfy certain desired criteria, where the stability is the most basic and fundamental requirement for the feedback system to be designed in order that the overall closed-loop configuration may work well. As discussed above, the input-output stability is the one with which we are mostly concerned. In this case, the question will be how to design an S_2 for the given S_1, such that all the conditions stated in Theorem 3.11 or Corollary 3.12 are satisfied.

To be more precise, let us consider the feedback system in Figure 3.5, which is a special case of the general system shown in Figure 3.4, with $S_1 = C, S_2 = P, e_1 = e, y_2 = y, u_1 = u$, and $u_2 = 0$ therein; where P is the given plant, a nonlinear operator in general; C the compensator (another nonlinear operator) to be designed; u the vector-valued reference control input; y the vector-valued system output; and e the vector-valued error signal between the input u and the output y through the unity feedback.

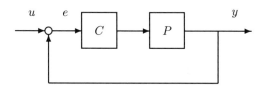

Figure 3.5. A nonlinear system with unity feedback.

Let X^e be the extended linear space associated with a given Banach space X, to which u, e and y belong, and let $U, V \subseteq X^e$ be two linear subspaces, namely:

$$U \subseteq X^e \text{ where } U = \{ u : \quad \|u_T\|_X < \infty \text{ for all } T < \infty \} \quad (3.11)$$

and

$$V \subseteq X^e \text{ where } V = \{ v : \quad \|v_T\|_X < \infty \text{ for all } T < \infty \}. \quad (3.12)$$

U and V will be used respectively as the admissible class of control inputs and the allowable maximum dynamic range for the outputs. Moreover, let

$$D = \{ e : \quad e = u - y, \quad u \in U, \ y \in V \}, \qquad (3.13)$$

which is also a linear subspace of X^e with $\|e_T\|_X < \infty$ for all $T \in [0, \infty)$. In designing the compensator C below, we will restrict the domain of C to be the linear subspace D.

From Figure 3.5, we see that the three signals u, y, e are connected together, so that linear operations must be allowed to be performed among them. Since these three signals (and their linear combinations) belong to three linear spaces, respectively, we see that the most natural and easiest way to handle this is to let $D = U = V$ in the closed-loop configuration Figure 3.5. We will impose this assumption in the following.

Now, suppose that the plant operator P is given, which may not be input-output stable. We first introduce the admissible class $\mathcal{S}$ of nonlinear compensators C. Let $Lip(D) = Lip(D, D)$ denote the family of nonlinear generalized Lipschitz operators that map D to itself. It was shown in Corollary 3.4 that all generalized Lipschitz operators are causal when the extended linear space setting is considered. We define $\mathcal{S}$ to be a subset of $Lip(D)$ such that the composite operator $PC(\cdot) := P \circ C(\cdot)$ is also a member of $Lip(D)$, namely:

$$\mathcal{S} = \{ C \in Lip(D) : \quad PC \in Lip(D) \}. \qquad (3.14)$$

Here, we do not have to require the plant operator P to be input-output stable, Lipschitz, or bounded in any sense, provided that the composite operator PC behaves as a generalized Lipschitz operator. An example of this case can be found in Example 3.15 given below. If P is a bounded operator in a certain operator norm, for example, if

$$\|P\| := \|P(0)\|_X + \sup_{\substack{x \in \mathcal{D}(P) \\ x \neq 0}} \frac{\|P(x) - P(0)\|_X}{\|x\|_X} < \infty,$$

such that it also satisfies $\|PC\|_{Lip} < \infty$ (usually, we have $\|PC\|_{Lip} \leq \|P\| \ \|C\|_{Lip} < \infty$), then by Corollary 2.5, we know that the set $\mathcal{S}$ defined in (3.14) is an (infinite-dimensional) Banach space. This is

true in the extended linear space setting as remarked at the end
of Chapter 2, Section 4. Hence, $\mathcal{S}$ is a very large family. If P is
a generalized Lipschitz operator in $Lip(D)$, then by Theorem 2.6,
PC is certainly in $\mathcal{S}$, and in this special case we have $\mathcal{S} = Lip(D)$,
which is a near-ring with unity I (the identity operator defined on
D), again by Theorem 2.6.

Now we are in a position to establish the following important
result:

3.14. Theorem. *Let $\mathcal{S}$ be the admissible class of nonlinear compensators C defined by 3.14 for the feedback system shown in Figure
3.5. Then, the subset $\mathcal{S}_0$ of $\mathcal{S}$ defined below consists of all the compensators C that stabilize the overall feedback system:*

$$\mathcal{S}_0 = \{\, C \in \mathcal{S} : \quad \|PC\| < 1 \,\}, \tag{3.15}$$

where $\|\cdot\|$ is the Lipschitz semi-norm defined as in (2.16).

Proof. It follows from Figure 3.5 that

$$e + PC(e) = u,$$

which is a vector-valued nonlinear equation in the error signal e for
each fixed input u. The condition in (3.15) ensures that the composite operator PC is Lipschitz with the norm strictly less than 1
uniformly on bounded subsets of D. By Theorem 2.15, PC is a contraction mapping from D into itself, so that for each $u \in D$, there
is a unique $e \in D$ satisfying the above equation. This implies that
the overall feedback system shown in Figure 3.5 is input-output
stable. $\square$

It is important to note that this theorem holds for any subset
D (need not be bounded nor be a linear space) provided that the
feedback system shown in Figure 3.5 is well connected and the
Lipschitz semi-norm $\|PC\|$ and norm $\|PC\|_{Lip}$ are both well defined
on D. In this case, of course, the set $\mathcal{S}$ defined in (3.14) is not a
Banach space in general, although D is an extended linear space.

We also remark that the condition in (3.15) is simultaneously
imposed on both P and C. Hence, even if P is arbitrarily given,

the feedback system can be stabilized provided that a compensator C satisfying the condition $\|PC\| < 1$ can be found. To see this, let us consider the following simple mathematical example, where for simplicity we only consider a standard Banach space.

3.15. Example. Let $D = L_\infty([0,\infty), R)$. Then, the operator P defined on D by

$$P(e)(t) := 1 + e(t) + e^2(t) + e^3(t) + \cdots$$

is not even bounded on D. However, with the operator C defined by

$$C(e)(t) := \begin{cases} \dfrac{1}{3}e(t) & \text{for } \|e\|_{L_\infty} \leq 1 \\[2ex] \dfrac{1}{3} & \text{for } \|e\|_{L_\infty} > 1, \end{cases}$$

we have

$$\|PC\|_{Lip} := 1 + \sup_{\substack{e_1,e_2\in D \\ e_1\neq e_2}} \frac{\|PC(e_1) - PC(e_2)\|_{L_\infty}}{\|e_1 - e_2\|_{L_\infty}}$$

$$\leq 1 + \frac{1}{3} + \frac{2}{3^2} + \frac{3}{3^3} + \cdots < \infty$$

on D. Hence, $PC \in Lip(D)$. Moreover, C is input-output stable from D to D and stabilizes P in the sense that $PC : D \to D$.

It is evident that the condition $\|PC\| < 1$ stated in Theorem 3.14 is in general only a sufficient condition for the finite-gain input-output stability of the feedback system. However, for such a general nonlinear setting, it does not seem to be easy to derive a necessary and sufficient condition. From a practical point of view, since S_0 is an open set (containing an open ball) in the infinite-dimensional Banach space S, it is usually large enough for the purpose of the design of stabilizing feedback compensators.

Next, let us consider a more complicated case, namely, the nonlinear feedback system shown in Figure 3.6, which is the same as that shown in Figure 3.4, with $S_1 = P$ (the given plant operator) and $S_2 = C$ (the compensator to be designed).

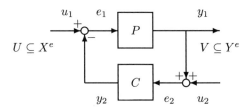

Figure 3.6. A nonlinear feedback system.

Let X^e be an extended linear space to which u_1, y_2, and e_1 belong, and Y^e another extended linear space to which u_2, y_1, and e_2 belong. Let

$$U \subseteq X^e \text{ where } U = \{ u : \quad \|u_T\|_X < \infty \text{ for all } T < \infty \} \quad (3.16)$$

be the admissible class of control inputs and

$$V \subseteq Y^e \text{ where } V = \{ v : \quad \|v_T\|_Y < \infty \text{ for all } T < \infty \} \quad (3.17)$$

be the allowable range of system outputs, as shown in Figure 3.6. For the finite-gain input-output stability, we will require here that $u_1, e_1, y_2 \in U$ and $u_2, e_2, y_1 \in V$. For the given nonlinear plant operator $P : X^e \to Y^e$, our aim is again to design a compensator C such that the overall closed-loop configuration shown in Figure 3.6 is (finite-gain input-output) stable. For some practical purposes, such as the so-called robust stabilization to be studied later, we introduce a sup-norm for the composite operator of P and C by defining

$$\|P(* - C(\cdot))\| := \|P(-C(0))\|_Y + \sup_{\substack{T \in [0,\infty) \\ u \neq 0}} \sup_{\substack{u \in U \quad e_1, e_2 \in V \\ [e_1]_T \neq [e_2]_T}}$$

$$\frac{\|[P(u_T - [C([e_1]_T)]_T)]_T - [P(u_T - [C([e_2]_T)]_T)]_T\|_Y}{\|u_T\|_X + \|[e_1]_T - [e_2]_T\|_Y}. \quad (3.18)$$

This is a mixed two-variable operator norm, which is actually the generalized Lipschitz norm for fixed $u \in U$ or fixed $e_1, e_2 \in V$. Let

$$\mathcal{S} = \{ C \in Lip(V, U) : \quad \|P(* - C(\cdot)\| < \infty \} \quad (3.19)$$

be the admissible class of nonlinear compensators C. Then, since in Figure 3.6 we have

$$e_2 = u_2 + P(e_1) = u_2 + P(u_1 - C(e_2)),$$

the nonlinear (two-variable) mapping

$$P(* - C(\cdot)): \quad (u_1, u_2) \to e_2$$

is a contraction if its norm $\|P(* - C(\cdot))\|$ is strictly less than 1. Consequently, if $u_1 \in U, u_2 \in V$, and if $y_1 = e_2 - u_2 \in V$ for all $e_2, u_2 \in V$, then the overall system is input-output stable in the sense that the system maps the two inputs $u_1 \in U$ and $u_2 \in V$ to the output $y_1 \in V$, namely, $S((U, V)) \subseteq V$. Hence, similar to Theorem 3.14, we have the following result:

3.16. Theorem. *Let S be the admissible class of nonlinear compensators C defined by (3.19) for the feedback system shown in Figure 3.6. Then, the subset S_0 of S defined below consists of compensators C that stabilize the overall system:*

$$S_0 = \{\, C \in S: \quad \|P(* - C(\cdot))\| < 1 \,\}. \tag{3.20}$$

Note that the input-output stability under consideration here is actually of finite gain.

§ 5. Relationship Between Input-Output and Lyapunov Stabilities

In this section, we study briefly the relationship between the input-output stability of a nonlinear feedback system and the Lyapunov stability of a related nonlinear control system described by the standard state-variable representation, where the latter has been widely studied in connection with nonlinear ordinary differential equations and the mathematical optimal control theory.

Consider, therefore, a nonlinear system described by the following first-order vector-valued ordinary differential equation:

$$\begin{cases} \dot{x}(t) = Ax(t) - f\big(x(t), t\big) \\ x(0) = x_0 \,, \end{cases} \tag{3.21}$$

with an equilibrium solution $\bar{x}(t) = 0$, where $x(t)$ is an $n \times 1$ real vector-valued measurable function of the time variable $t \in [0, \infty)$, the so-called state-vector, A is an $n \times n$ constant matrix whose eigenvalues are assumed to have negative real parts, and $f : R^n \times R^1 \to R^n$ is a real vector-valued integrable nonlinear function of $t \in [0, \infty)$. By adding and then subtracting the term $Ax(t)$, a general nonlinear system can always be written in this form.

To establish a connection between this system and the closed-loop feedback configuration that we have studied in some depth in the previous sections, we define

$$\begin{cases} x(t) = u(t) - \displaystyle\int_0^t e^{(t-\tau)A} y(\tau) d\tau \\ y(t) = f(x(t), t) \end{cases} \tag{3.22}$$

with $u(t) = x_0 e^{tA}$; then we can implement system (3.21) by a feedback configuration as depicted in Figure 3.7, where the error signal $e(t) = x(t)$, the plant $P(\cdot)(t) = f(\cdot, t)$, and the compensator $C(\cdot)(t) = \int_0^t e^{(t-\tau)A}(\cdot)(\tau)d\tau$.

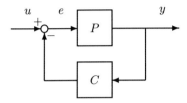

Figure 3.7. A nonlinear feedback system.

Now, denoting respectively by $L_p([0,\infty), R^n)$ the standard L_p space of $x(t) \in R^n$ and by $L_p^e([0,\infty), R^n)$ the n-dimensional extended L_p space defined on $[0,\infty)$, namely, the extended linear space of all $x(t) \in R^n$ satisfying $\|x_T\|_{L_p} < \infty$ for all $T \in [0,\infty)$, we can establish the following result:

3.17. Theorem. *Consider the nonlinear system (3.21) and its associated feedback configuration shown in Figure 3.7. Suppose that $U = V = L_p([0,\infty), R^n) \subseteq L_p^e([0,\infty), R^n)$, where $1 < p < \infty$. Then, if the feedback system shown in Figure 3.7 is input-output stable from U to V, the nonlinear system (3.21) is globally asymptotically stable above the equilibrium point 0 in the sense of Lyapunov, namely: $|x(t)| \to 0$ as $t \to \infty$, where $|\cdot|$ is the Euclidean norm of the vector.*

Proof. Since all eigenvalues of the constant matrix A have negative real parts, we have

$$\left|x_0 e^{tA}\right| \le M e^{-\alpha t}$$

for some $0 < \alpha, M < \infty$ for all $t \in [0,\infty)$, so that $|u(t)| = |x_0 e^{tA}| \to 0$ as $t \to \infty$. Hence, in view of the first equation of (3.22), if we can prove that

$$v(t) := \int_0^t e^{(t-\tau)A} y(\tau) d\tau \to 0 \qquad \text{as} \quad t \to \infty$$

in the Euclidean norm, then it will follow that

$$|x(t)| = |u(t) - v(t)| \to 0 \qquad \text{as } t \to \infty.$$

Write

$$v(t) = \int_0^{t/2} e^{(t-\tau)A} y(\tau)d\tau + \int_{t/2}^t e^{(t-\tau)A} y(\tau)d\tau$$

$$= \int_{t/2}^t e^{\tau A} y(t-\tau)d\tau + \int_{t/2}^t e^{(t-\tau)A} y(\tau)d\tau.$$

Then, by the Hölder inequality we have

$$|v(t)| \le \left| \int_{t/2}^t e^{\tau A} y(t-\tau)d\tau \right| + \left| \int_{t/2}^t e^{(t-\tau)A} y(\tau)d\tau \right|$$

$$\le \left(\int_{t/2}^t |e^{\tau A}|^q d\tau \right)^{1/q} \left(\int_{t/2}^t |y(t-\tau)|^p d\tau \right)^{1/p}$$

$$+ \left(\int_{t/2}^t |e^{(t-\tau)A}|^q d\tau \right)^{1/q} \left(\int_{t/2}^t |y(\tau)|^p d\tau \right)^{1/p}$$

$$\le \left(\int_{t/2}^\infty |e^{\tau A}|^q d\tau \right)^{1/q} \left(\int_0^\infty |y(\tau)|^p d\tau \right)^{1/p}$$

$$+ \left(\int_0^\infty |e^{\tau A}|^q d\tau \right)^{1/q} \left(\int_{t/2}^\infty |y(\tau)|^p d\tau \right)^{1/p}.$$

Since all eigenvalues of A have negative real parts and since the feedback system is input-output stable from U to V, so that $y \in V = L_p([0,\infty), R^n)$, we have

$$\lim_{t\to\infty} \int_{t/2}^\infty |e^{\tau A}|^q d\tau = 0$$

and

$$\lim_{t\to\infty} \int_{t/2}^\infty |y(\tau)|^p d\tau = 0.$$

Hence, it follows that $|v(t)| \to 0$ as $t \to \infty$. $\qquad\qquad\square$

Exercises

3.1. Verify that the feedback configuration shown in Figure 3.2 is a special case of the one shown in Figure 3.1.

3.2. Verify that the nonlinear integral operator S defined in Example 3.2 is causal.

3.3. Let X^e be the extended linear space associated with a given Banach space X of complex-valued measurable functions defined on the time domain $[0, \infty)$. Verify that for $X = L_p, 1 \leq p \leq \infty$, the norm $\|f_T\|_X$ is monotonically increasing as $T \to \infty$ for all $f \in X^e$ and satisfies $\|f_T\|_X \to \|f\|_X$ as $T \to \infty$ for all $f \in X$.

3.4. Consider the nonlinear feedback system shown in Figure 3.1, with the two nonlinear subsystems defined by

$$S_1 : e_1(t) \to \frac{e^{-t}}{2} \sin[e_1(t)] \quad S_2 : e_2(t) \to \frac{e^{-t}}{2} + \sin[e_1(t)].$$

Determine the possible upper bounds for $\|[e_1]_T\|_{L_\infty}$ and $\|[e_2]_T\|_{L_\infty}$ using the small gain theorem.

3.5. Consider the nonlinear control system shown in Figure 3.3. Give examples to show that the same control system S can be input-output stable under certain norms but unstable under some others, even for the same set of control input signals.

3.6. Give examples to show that when two individually input-output stable subsystems S_1 and S_2 are connected together in the closed-loop configuration shown in Figure 3.4, the resulting overall feedback system may be jointly input-output unstable.

3.7. Complete the discussion on Example 3.13 by supplying detailed verifications.

3.8. Verify that the two nonlinear systems described by (3.21) and (3.22), respectively, are equivalent. Is Theorem 3.17 valid for $p = 1$ and $p = \infty$? Prove or disprove your claim.

References

[1] G. Chen and R.J.P. de Figueiredo, "Optimal nonlinear feed-
    back system design for a general tracking problem," in *Mathe-
    matical Theory of Networks and Systems*, (M.A. Kaashoek *et
    al*, eds.), Birkhäser, Boston, 1989, 429–436.

[2] G. Chen and R.J.P. de Figueiredo, "On robust stabilization
    of nonlinear control systems," *Sys. Contr. Letts.*, **12** (1989),
    373–379.

[3] C.K. Chui and G. Chen, *Linear Systems and Optimal Control*,
    Springer-Verlag, New York, 1989.

[4] C.K. Chui and G. Chen, *Signal Processing and Systems The-
    ory: Selected Topics*, Springer-Verlag, New York, 1992.

[5] C.A. Desoer and M. Vidyasagar, *Feedback Systems: Input-
    Output Properties*, Academic Press, New York, 1975.

[6] R.J.P.de Figueiredo and G. Chen, "Optimal disturbance re-
    jection for nonlinear control systems," *IEEE Trans. Auto.
    Contr.*, **12** (1989), 1242–1248.

[7] L.T. Grujić and A.N. Michel, "Exponential stability and tra-
    jectory bounds of neural networks under structural variations,"
    IEEE Trans. Circ. Sys., **38** (1991), 1182-1192.

[8] H. Khalil, *Nonlinear Systems*, Macmillan, New York, 1992.

[9] M. Vidyasagar, *Nonlinear Systems Analysis*, Prentice-Hall,
    Englewood Cliffs, N. J., 1978; 2nd Ed., 1993.

[10] J.C. Willems, *The Analysis of Feedback Systems*, MIT Press,
    Cambridge, MA, 1971.

Chapter 4

Optimal Design of Nonlinear Feedback Control Systems

This chapter is devoted to the investigation of some optimal design strategies for nonlinear feedback control systems. Based on the developments in the previous chapters, we are now in a position to study some general theories and methods for optimal compensator design, at least qualitatively, for multi-input/multi-output nonlinear feedback control systems. Problems under consideration will be formulated in a general extended linear space setting in the time domain. In this chapter, we investigate three typical feedback design problems that have been frequently discussed from different viewpoints in the nonlinear systems control literature. In the first section, we provide a brief motivation for these problems, by consideration of a single-link robot arm. In Section 2, we discuss in detail an optimal tracking problem. We first give a precise description and then a mathematical formulation for the problem. For the optimal tracking problem, we provide basic results, such as the solvability, and discuss briefly a general constructive procedure for finding an optimal solution. Then, we study very briefly, in Section 3, an optimal disturbance rejection problem, and in Section 4, a robust stabilization problem, based on similar ideas.

§ 1. Motivation for Optimal Feedback Controller Design

In this section, we first discuss some practical engineering design problems for a simple physical model, a single-link robot arm, as motivation for the general optimal design problems to be studied in this chapter.

Consider a single-link robot arm as shown in Figure 4.1, which has a simple mathematical description

$$J\ddot{\theta}(t) + MgL\sin\theta(t) = \tau(t), \tag{4.1}$$

where $\theta(t)$ is the rotational angle of the link with the vertical line as reference, L and M the (half) length and (concentrated) mass of the link, respectively, J the rotational inertia about the joint-axis, g the gravity acceleration constant, and $\tau(t)$ the control torque input to the link from the motor.

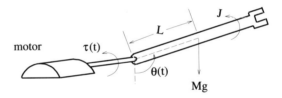

Figure 4.1. A single-link robot arm.

Suppose that we want to find the control torque input $\tau(t)$ that can drive the link angle $\theta(t)$ from its initial position, say $\theta_0 = 0^o$ with angular velocity $\dot{\theta}_0 = 0$, to a desired (constant) angle θ_d. Defining

$$e(t) = \theta(t) - \theta_d$$
$$\dot{e}(t) = \dot{\theta}(t) - \dot{\theta}_d = \dot{\theta}(t),$$

for the tracking errors, and using them to drive a controller (to be designed) such that the controller can create a suitable torque $\tau(t)$ to drive the motor for the task, we obtain a standard feedback control system as shown in Figure 4.2.

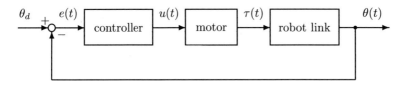

Figure 4.2. A feedback control system of the robot arm.

Suppose that we use a standard PD (proportional-derivative) controller of the form

$$u(t) = -K_d(\dot{\theta}(t) - \dot{\theta}_d) - K_p(\theta(t) - \theta_d)$$
$$= -K_d\dot{e}(t) - K_pe(t), \qquad (4.2)$$

where K_p and K_d are constant gains (proportional to the angular error and its derivative, respectively) to be determined. This PD controller is as shown in Figure 4.3.

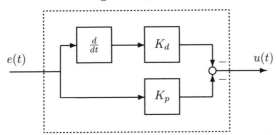

Figure 4.3. The structure of the PI controller.

Suppose, on the other hand, that the PD controller is designed to create a motor control torque $\tau(t)$ in such a way that

$$\tau(t) = Ju(t) + MgL\sin\theta(t), \qquad (4.3)$$

then by combining Eqs. (4.1) to (4.3), we arrive at the following dynamic equation:

$$J\ddot{\theta}(t) + MgL\sin\theta(t) = \tau(t)$$
$$= Ju(t) + MgL\sin\theta(t)$$
$$= J[-K_d\dot{e}(t) - K_pe(t)] + MgL\sin\theta(t),$$

so that

$$\begin{cases} \ddot{e}(t) + K_d \dot{e}(t) + K_p e(t) = 0 \\ e(0) = \theta_0 - \theta_d = -\theta_d, \quad \dot{e}(0) = 0. \end{cases}$$

Since its eigenvalues are

$$s_{1,2} = \frac{-K_d \pm \sqrt{K_d^2 - 4K_p}}{2},$$

we see that by choosing $K_d > 0$ we can guarantee that $e(t) \to 0$ as $t \to \infty$, and by further using $K_p = K_d^2/4$ we can zero out the imaginary part of the eigenvalues, so that the tracking performance can be greatly improved in the sense that the oscillations in the tracking process can be eliminated.

Now, let us assume that the above mathematical model does not describe correctly the actual physical robot arm, due to the ignorance of damping effect, friction, etc., and assume that the true model should be

$$J\ddot{\theta}(t) + f(t, \theta(t), \dot{\theta}(t)) = \tau(t) \tag{4.4}$$

for a nonlinear function $f(\cdot)$ that contains the well-modeled part $MgL \sin \theta(t)$ and a (small but unknown) misfit part denoted by $\Delta f(t, \theta(t), \dot{\theta}(t))$:

$$f(t, \theta(t), \dot{\theta}(t)) = MgL \sin \theta(t) + \Delta f(t, \theta(t), \dot{\theta}(t)).$$

The overall feedback control configuration is shown in Figure 4.4. In Figure 4.4, the external input

$$\xi(t) = J^{-1} \Delta f(t, \theta(t), \dot{\theta}(t)),$$

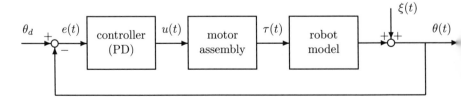

Figure 4.4. Overall Feedback Control System of the Robot Arm

where to handle the model-misfit uncertainty, we consider $\xi(t)$ to be a disturbance and will design an appropriate controller, usually different from the PD controller discussed above and preferably optimal in some sense, to reject the disturbance while achieving the goal of tracking:

$$e(t) \to 0 \qquad \text{as} \quad t \to \infty.$$

Indeed, since we actually have the model in Fig. 4.4 while using the control design in Fig. 4.2, it follows from Eqs. (4.2) and (4.4) that

$$\begin{cases} \ddot{e}(t) + K_d\dot{e}(t) + K_pe(t) = -\xi(t) \\ e(0) = -\theta_d, \quad \dot{\theta}(0) = 0, \end{cases} \tag{4.5}$$

which yields the expected tracking result, provided that the unknown disturbance $\xi(t)$ is bounded and the PI control gains are chosen appropriately.

Here, the approach is not to correct the inexact mathematical modeling since this is usually impossible for many practical systems where some quantities (damping, friction, etc.) cannot be accurately modeled. Instead, the idea in the above formulation is to use the inexact mathematical model to perform the control, which is practical. As long as the designed controller can ensure the tracking error to decay, $e(t) \to 0$ as $t \to \infty$, the task is done.

In the above, we have discussed some features of the tracking and disturbance rejection problems. Another important problem, involving stability, has traditionally been of great interest in control system design and will be addressed in Section 4.

§ 2. Optimal Tracking

In this section, we consider a general multi-input/multi-output (MIMO) nonlinear feedback system defined in an extended linear space setting in the time domain as shown in Figure 4.5. The system equations are given by

$$\begin{cases} e = r - y \\ y = PC(e) + W_2(d) \\ r = W_1(u)\,, \end{cases} \qquad (4.6)$$

where r denotes the reference input signal modeled as the output of a linear filter W_1 driven by an external source u, e the error signal between the reference signal r and the system output y in which y is required to follow the given reference signal r, v the possible disturbance input modeled as the output of another linear filter W_2 driven by an external source (noise) d, in which W_2 equals zero in case no disturbances are considered, and P and C are respectively the plant and compensator operators, which are nonlinear in general.

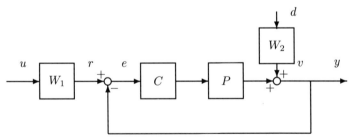

Figure 4.5.   The nonlinear feedback tracking system.

Let X^e, Y^e, and Z^e be three extended linear spaces of l-, p-, and q-dimensional complex-valued measurable functions defined on the time domain $[0, T)$ where $T \le \infty$ and $1 \le l, p, q < \infty$, such that $u \in X^e$; $y, r, e \in Y^e$; and $d \in Z^e$. Let $U \subseteq X^e$ and $V \subseteq Y^e$ be the input and output spaces, respectively, and let $\mathcal{D}(\cdot)$ and $\mathcal{R}(\cdot)$ denote respectively domains and ranges for operators.

Recall that if we want the system under consideration to be input-output stable, then for any input in U the corresponding system output must be in V.

We first clarify some necessary assumptions imposed on this feedback system for its well-posedness. Assume that $W_1 : X^e \to Y^e$ and $W_2 : Z^e \to Y^e$ are both bounded linear operators and $P : \mathcal{R}(C) \to Y^e$ and $C : Y^e \to \mathcal{D}(P)$ are two nonlinear operators on their domains in which P is given and C is to be designed. Note that even if both P and C are bounded (in certain operator norms) and continuous, their boundedness and continuity do not imply the finite-gain input-output stability of the overall feedback system because of the closed-loop configuration that may map a signal to somewhere outside the output space. For the consistency of the overall feedback system, we will assume $\mathcal{R}(C) \subset \mathcal{D}(P)$ and $\mathcal{R}(W_1) \dotplus (\mathcal{R}(W_2) \dotplus \mathcal{R}(P)) \subseteq \mathcal{D}(C)$, where $\dotplus$ is the geometric sum of two sets of vectors.

In the following study of this closed-loop configuration, the plant P (a nonlinear operator) as well as the two filters W_1 and W_2 (two linear operators) is assumed to be given and the compensator C (another nonlinear operator) is to be designed. Criteria for the design will be further discussed later.

Before describing the optimal tracking design problem more precisely, we need to clarify the admissible classes of source inputs and disturbances. As mentioned above, we consider two linear subspaces as input and output spaces, namely:

$$U \subseteq X^e \text{ where } U = \{u : \quad \|u_T\|_X < \infty \text{ for all } T \in [0, \infty)\} \quad (4.7)$$

and

$$V \subseteq Y^e \text{ where } V = \{v : \quad \|v_T\|_Y < \infty \text{ for all } T \in [0, \infty)\} \quad (4.8)$$

and will choose the admissible class of disturbances to be a subspace D in Z^e defined by

$$D \subseteq Z^e \text{ where } D = \{d : \quad \|d_T\|_Z < \infty \text{ for all } T \in [0, \infty)\}, \quad (4.9)$$

where $[\cdot]_T$ is the truncation of the function with respect to T as defined before in Chapter 2. Since the plant outputs y will be required to belong to the output space $V \subseteq Y^e$ in the consideration of input-output stability of the system, for the well connection of the feedback control system shown in Figure 4.5 above, we will simply require that $r \in Y^e$, so that $e = r - y \in Y^e$.

Now, the optimal tracking design problem can be stated as follows.

4.1. Problem. *Given P, W_1, W_2, D, U, and V as described above. The problem is to design a compensator C belonging to certain admissible class $\mathcal{S}$ so as to stabilize the overall feedback system while minimizing the norm of the error signal e between the system output y and the reference signal r, where the stability is in the sense of input-output, namely, for any input $u \in U$, we have the corresponding output $y \in V$.*

The physical meaning of this problem is simply that we want to stabilize the feedback system while controlling the system output y to follow the reference input signal r optimally in the sense that the error $e = r - y$ is the smallest possible in the norm of the space to which e belongs.

Throughout this section, the stability under consideration is of input-output as described in the problem statement above.

To formulate this problem mathematically, we first have to specify the admissible class $\mathcal{S}$ to which the compensator C belongs. For this purpose, we recall the notion of generalized nonlinear Lipschitz operators studied in Chapter 2 and the input-output stability studied in Chapter 3. Recall from Corollaries 3.4 and 3.5 that a generalized Lipschitz operator is always causal and guarantees the uniqueness of the output signal when the underlying framework is in an extended linear space setting. Hence, we formulate the above optimal tracking problem using generalized Lipschitz operators.

Let $Lip(V)$ denote the family of generalized Lipschitz operators mapping from V to itself, where V is the output space for the system as well as the "input space" for the compensator C defined in (4.3). Then, from Theorem 2.21 we know that $Lip(V)$ is an infinite-dimensional Banach space containing all bounded linear operators defined on V, and is hence a very large family. Moreover, it follows from Corollary 2.5 and Theorem 2.21 that for the given plant operator P, the family

$$\mathcal{S} = \{C \in Lip(V) : \quad PC \in Lip(V)\} \qquad (4.10)$$

is an infinite-dimensional Banach space as well, as was pointed out at the end of Chapter 2, Section 4.

Hence, using S defined in (4.10) as an underlying operator space for the admissible class of compensators, we have a very large framework for the design purposes. The advantages of choosing this Banach space, S, to be the underlying space for compensators will be further confirmed by the following two results:

4.2. Theorem. *Let P and S be given as described above. Then, the subset S_0 of S defined below consists of compensators C that stabilize the overall feedback system* (4.6):

$$S_0 = \{C \in S : \quad \|PC\| < 1\}, \qquad (4.11)$$

where $\|\cdot\|$ is the Lipschitz semi-norm of the operator.

This is just Theorem 3.14. Moreover, the following result has been established in Theorem 2.22.

4.3. Theorem. *Under the condition that $\|PC\| < 1$, the nonlinear operator*

$$I(\cdot) + PC(\cdot) : \quad V \to V$$

is invertible, and its inverse, denoted by $(I + PC)^{-1}$, is also a generalized Lipschitz operator in $Lip(V)$ satisfying

$$\|(I + PC)^{-1}\|_{Lip} \leq \|(I + PC)^{-1}(0)\|_Y + (1 - \|PC\|)^{-1}. \quad (4.12)$$

Hence, once we have chosen the family S_0 to be the admissible class of compensators to be designed, the overall feedback system (4.6) is stable provided that $W_1 : U \to V$, and is mathematically well connected in the sense that all the operators (except perhaps the given plant P) in the system, namely, C, PC, and $(I + PC)^{-1}$, belong to the same family $Lip(V)$, which is an infinite-dimensional Banach space, and also in the sense that the overall feedback system is causal with unique internal signals. As is well known, such a mathematical well-connection is very difficult to achieve for the general MIMO nonlinear feedback system shown in Figure 4.5 without using the generalized Lipschitz operators framework. This is one of the main advantages of the generalized Lipschitz operators theory approach.

Now, we are in a position to pose the optimal tracking design problem precisely in a mathematical manner.

Consider, therefore, relationships (4.6) and assume that all the conditions stated above are satisfied. It is then easy to see that

$$e + PC(e) = W_1(u) - W_2(d)$$

or, by Theorem 4.3,

$$e = (I + PC)^{-1}(W_1 u - W_2 d). \tag{4.13}$$

Recall that for an optimal tracking problem our purpose is to minimize (the norm of) the error signal e, where $e = r - y$ is the difference between the reference input r and the system output y, and y is required to follow r, while the overall feedback system must maintain to be (input-output) stable. Hence, observing that in the relationship (4.13) the two external inputs u and d are independent and that the norm of the error signal e should be minimized uniformly over all possible inputs u and d, we may formulate the problem as follows.

4.4. Problem.

$$\min_{C \in \mathcal{S}_0} \left\| (I + PC)^{-1} \right\|_{Lip},$$

where

$$\mathcal{S}_0 = \{ C \in \mathcal{S} : \quad \|PC\| < 1 \}.$$

Next, we consider solving Problem 4.4. We first point out that it can be verified by definition that the Lipschitz operator $(I + PC)^{-1}$ depends continuously on C (and so does the objective functional $\|(I + PC)^{-1}\|_{Lip}$), so that all the discussions and results given below can be directly applied.

We establish, at the outset, an existence theorem for optimal solutions to Problem 4.4. Note that although the admissible class of compensators, namely, the family $\mathcal{S}$ defined by (4.10), is an infinite-dimensional Banach space, the subset $\mathcal{S}_0$ given in (4.11) consisting of stabilizing compensators is only an open set (containing an open ball) in $\mathcal{S}$, which is neither compact nor closed. Hence, establishing an existence theorem for optimal solutions to the problem in such a general setting is difficult. However, for a continuous functional of the form $f(PC)$ in variable C on a (closed) compact set, the solvability becomes trivial. The following results are immediate.

4.5. Theorem. *For a continuous functional of the form $f(PC)$ in the variable $C \in S_0$, the minimization problem*

$$\min_{C \in S_0} f(PC)$$

is always solvable under one of the following conditions:
(i) *Let S^c be a compact set of S and let*

$$S_0^c = \left\{ C \in S^c : \quad \|PC\| < 1 \right\}.$$

Then, Problem 4.4 has a solution in S_0^c.
(ii) *Let S^b be a bounded subset in any finite-dimensional subspace of S and let*

$$S_0^b = \left\{ C \in S^b : \quad \|PC\| \le 1 - \epsilon \right\}$$

where $\epsilon > 0$ arbitrarily small. Then, Problem 4.4 has a solution in S_0^b.

It amounts to noting a fact from functional analysis that a continuous functional assumes its infimum on any compact sets in its domain. Hence, since the set S_0^c is compact, and since the set S_0^b is closed and bounded in a finite-dimensional space and hence is also compact, the above conclusions follow immediately.

We remark that because the polynomic operators or nonlinear Volterra operators of finite degrees that we studied in Chapter 1 constitute a dense set in the Banach space S of Lipschitz operators defined in (4.10), the following result is immediate. The result states that the infimum of the objective functional $f(PC)$ can be approached as close as possible. Hence, from the practical point of view, this observation is useful.

4.6. Corollary. *Let*

$$\delta = \inf_{C \in \mathcal{S}_0} f(PC)$$

and

$$\delta_k = \inf_{C_k \in \mathcal{S}_k} f(PC_k),$$

where $\mathcal{S}_k$ is either $\mathcal{S}_0^c$ or $\mathcal{S}_0^b$ defined respectively in (i) *or* (ii) *of Theorem 4.5 in a k-dimensional subspace of $\mathcal{S}$. Then, we have*

$$\delta_k \to \delta \qquad \text{as} \quad k \to \infty.$$

There are many nontrivial examples of (generalized) Lipschitz operators, the totality of which consists of a compact set in the infinite-dimensional space $\mathcal{S}$ defined by (4.5). The following is a simple mathematical example for the purpose of illustration.

4.7. Example. Let the admissible class $\mathcal{S}$ be the family of compensators having a uniformly convergent expansion of the form

$$C(v)(t) = c_0(v(t), t) + c_1(v(t), t) + c_2(v(t), t) + \cdots$$

for some nonlinear operators $c_k(\cdot, t)$ satisfying the (generalized) Lipschitz condition on V, where V is a linear subspace. Recall from functional analysis that all compact metric spaces are separable. Let $\{a_{nk}\}$ be the set of the so-called majorizing sequences of real numbers such that

$$\sum_{k=0}^{\infty} a_{nk} < \infty \quad \text{and} \quad \sum_{n=0}^{\infty} a_{nk} < \infty$$

for all $n, k = 0, 1, \ldots$, and let $\mathcal{S}_c$ be the family of nonlinear compensators defined by

$$\mathcal{S}_c = \left\{ C_n(\cdot)(t) = \sum_{k=0}^{\infty} c_{nk}(\cdot, t) \in \mathcal{S} : \quad n = 0, 1, \ldots \right.$$

$$\left. \sup_t |c_{nk}(v(t), t)| \le \frac{M^k}{k!} a_{nk}, \quad n, k = 0, 1, \ldots \right\},$$

for some fixed $M < \infty$ and for all $v \in V$. Then, $\mathcal{S}_c$ is an infinite-dimensional compact subspace of $\mathcal{S}$. Indeed, for all $v \in V$ and for each n, we have

$$\|C_n(v)\|_Y \leq \sum_{k=0}^{\infty} \sup_t |c_{nk}(v(t), t)| \leq \sum_{k=0}^{\infty} a_{nk} \sum_{k=0}^{\infty} \frac{M^k}{k!} < \infty.$$

Moreover, it is easily seen that every sequence in $\mathcal{S}_c$ has a convergent subsequence.

Technically, solving Problem 4.4 may be very difficult in general, since there is an inverse operator $(I + PC)^{-1}$ in the objective functional. Under a further technical assumption, however, we may simplify significantly the formulation and obtain a suboptimal solution for the problem. Specifically, if we assume that the compensator C to be designed is such that $(I + PC)^{-1}(0) = 0$ and observe that for any generalized Lipschitz operator $Q \in Lip(V)$ satisfying $PQ \in Lip(V)$ with $\|PQ\| < 1$, where again $\|\cdot\|$ is the Lipschitz semi-norm of the operator, $(I - PQ)^{-1}$ exists and is also in $Lip(V)$ by Theorem 4.3, then we may define

$$C := Q(I - PQ)^{-1}, \tag{4.14}$$

and let

$$\widetilde{S}_0 = \left\{ Q \in \mathcal{S}: \quad (I - PQ)^{-1}(0) = 0 \quad \text{and} \quad \|PQ\| < \frac{1}{2} \right\}, \tag{4.15}$$

then we can show that if Q^* is an optimal solution of the minimization

$$\min_{Q \in \widetilde{S}_0} \|I - PQ\|_{Lip},$$

then $C^* = Q^*(I - PQ^*)^{-1}$ is an optimal solution of Problem 4.4. First, we observe that if $Q \in \widetilde{S}_0$, then we have $\|(I - PQ)^{-1}\| < 2$ by (4.12), so that (4.14) yields

$$\|PC\| = \|PQ(I - PQ)^{-1}\| \leq \|PQ\| \, \|(I - PQ)^{-1}\| < 1,$$

which implies that $C \in \mathcal{S}_0$. Then, we observe that

$$\begin{aligned} I + PC &= I + PQ(I - PQ)^{-1} \\ &= [(I - PQ) + PQ](I - PQ)^{-1} \\ &= (I - PQ)^{-1}, \end{aligned}$$

or

$$(I + PC)^{-1} = I - PQ.$$

Hence, if we can find a Q^* such that $\|I - PQ^*\| \leq \|I - PQ\|$, then the corresponding C^* will satisfy $\|(I + PC^*)^{-1}\| \leq \|(I + PC)^{-1}\|$. This implies that instead of solving Problem 4.4, we may consider solving the following simpler one:

4.8. Problem.

$$\min_{Q \in \widetilde{\mathcal{S}}_0} \|I - PQ\|_{Lip},$$

where $\widetilde{\mathcal{S}}_0$ is defined in (4.15).

Note that Problem 4.8 is not equivalent to Problem 4.4 in a general nonlinear setting, since when Q runs over the set $\widetilde{\mathcal{S}}_0$, the corresponding C defined in (4.14) may only run over a subset of $\mathcal{S}_0$. Hence, an optimal solution for Problem 4.8 yields only a suboptimal solution for Problem 4.4 in general.

Once an optimal solution $Q^* \in \widetilde{\mathcal{S}}_0$ has been found, we obtain a suboptimal solution for the Problem 4.4 via

$$C^* = Q^*(I - PQ^*)^{-1}, \qquad (4.16)$$

which may be implemented as shown in Figure 4.6.

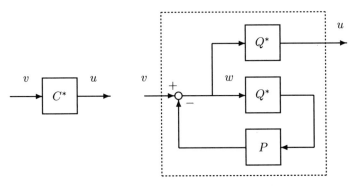

Figure 4.6. Two equivalent control systems.

As a matter of fact, for any input-output pairs (u, v) such that $v = C^*(u)$, we can verify that $v = Q^*(I - PQ^*)^{-1}(u)$, so that we have (4.16) and that the two blocks shown in Figure 4.6 are equivalent, in the sense that the same input produces the same output through the two systems. To see this, it amounts to noting that we have $w = u + PQ^*(w)$ in Figure 4.6, so that $w = (I - PQ^*)^{-1}(u)$, and hence

$$v = Q^*(w) = Q^*(I - PQ^*)^{-1}(u).$$

Finally in this section, we remark that the problem of solving the minimization 4.8 for an optimal solution $Q^* \in \widetilde{S}_0$ will become a nonlinear programming in general. Hence, efficient numerical techniques are needed. This can be seen, for example, from the case where the input control functions are restricted in a bounded subset of the input space V in the extended linear space Y^e and the admissible class S consists of compensators of the form

$$Q(\cdot) = q_0(t) + q_1(t)(\cdot)(t) + q_2(t)(\cdot)^2(t) + \cdots + q_n(t)(\cdot)^n(t).$$

In this case, what we need in determining an optimal $Q^*(\cdot)$ for the problem is to obtain $n + 1$ functions $q_k(t), k = 0, 1, \ldots, n$, in a certain function space, depending on the specific problem under consideration. This may be achieved by some standard or modified nonlinear programming technique. Nevertheless, the discussions given in this section for an optimal tracking design posed in such a general setting have provided some theoretical insights and mathematical principles for the purpose of qualitative analysis and design.

§ 3. Optimal Disturbance Rejection

In this section, we discuss briefly an optimal disturbance rejection problem for a general MIMO nonlinear feedback system defined in an extended linear space setting in the time domain. The system under consideration is the same as that shown in Figure 4.5.

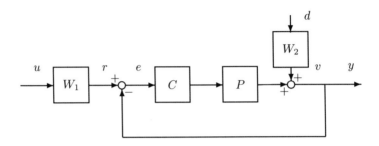

Figure 4.7.   The nonlinear system with disturbance.

The system equations are given by

$$\begin{cases} e = r - y \\ y = PC(e) + W_2(d) \\ r = W_1(u), \end{cases} \qquad (4.17)$$

where again r denotes the reference input signal modeled as the output of a linear filter W_1 driven by an external source u, e the error signal between the reference r and the system output y, v the random disturbance modeled as the output of a linear filter W_2 driven by an external noise d, and P and C are respectively the plant and compensator operators, which are nonlinear in general.

Let X^e, Y^e, and Z^e be respectively three extended linear spaces of l-, p-, and q-dimensional real-valued measurable functions defined on the time domain $[0, T)$, where $T \leq \infty$ and $1 \leq l, p, q < \infty$. As shown in Figure 4.7, $u \in X^e$, $y, r, e, v \in Y^e$, and $d \in Z^e$.

With the notations $\mathcal{D}(\cdot)$ and $\mathcal{R}(\cdot)$ being the domain and range, respectively, of the argument operators, we assume that $W_1 : X^e \rightarrow$

Y^e and $W_2 : Z^e \to Y^e$ are both bounded linear operators such that $\mathcal{R}(W_1) \subseteq \mathcal{R}(W_2) = \mathcal{D}(C)$. The reason for the condition

$$\mathcal{R}(W_1) \subseteq \mathcal{R}(W_2)$$

will be clear later in the problem formulation, which may be understood here from the physical viewpoint as that the range of noise inputs is usually (much) larger than that of the signals, especially in the consideration of random disturbance rejection. Recall that even if $P : \mathcal{R}(C) \to Y^e$ and $C : Y^e \to \mathcal{D}(P)$ are both bounded and continuous nonlinear operators, it does not imply the input-output stability of the system because of the closed-loop configuration. Assume that $\mathcal{R}(C) \subseteq \mathcal{D}(P)$ and $\mathcal{R}(P) \subseteq \mathcal{R}(W_2) = \mathcal{D}(C)$. Under all these assumptions, the closed-loop configuration shown in Figure 4.7 is considered to be well connected. The stability and causality, as well as the uniqueness of internal signals, will be taken into account later, after we imposed the generalized Lipschitz condition on the corresponding nonlinear operators in the system, as we did in the previous section.

We next introduce the admissible classes of source inputs and random disturbances. Here in this section, we suppose that the class of random disturbances is a bounded subset in the corresponding extended linear space defined by

$$D = \left\{ d \in Z^e : \; \|d_T\|_X \leq M_d < \infty \text{ for all } T \in [0, \infty) \right\}. \quad (4.18)$$

The input and output spaces U and V, however, may be defined as before, namely,

$$U \subseteq X^e \text{ where } U = \left\{ u : \; \|u_T\|_X < \infty \text{ for all } T \in [0, \infty) \right\} \quad (4.19)$$

and

$$V \subseteq Y^e \text{ where } V = \left\{ v : \; \|v_T\|_Y < \infty \text{ for all } T \in [0, \infty) \right\}. \quad (4.20)$$

Here, it should be noticed that since the family of admissible disturbances is no longer a linear subspace, the actual systems outputs, y, constitute a subset in V that is not necessarily a linear subspace of Y^e.

Now, given P, W_1, W_2, U, D, and V, our objective is to solve the following problem.

4.9. Problem. *Design a compensator C belonging to a certain admissible class S so as to stabilize the system while minimizing the norm of the error e between the system output y and the reference signal r by reducing the effect of the random disturbance d on the system.*

In order to pose the problem mathematically, we need to introduce an admissible class S of nonlinear compensator operators. Recall that $Lip(V)$ denotes the family of generalized Lipschitz operators defined on V. Similar to (4.10), we define the admissible class of nonlinear compensators to be the following:

$$S = \{C \in Lip(V): \quad PC \in Lip(V)\}. \qquad (4.21)$$

Recall also that S is an infinite-dimensional Banach space and is hence a very large family.

As an analogue of Theorem 4.2, the subset S_0 of S defined below consists of compensators C that stabilize the system:

$$S_0 = \{C \in S: \quad \|PC\| < 1\}, \qquad (4.22)$$

where again $\|\cdot\|$ is the Lipschitz semi-norm of the operator.

Also, as an analogue of Theorem 4.3, under the condition $\|PC\| < 1$ we know that the nonlinear operator $I + PC$ is invertible on V, and moreover its inverse $(I + PC)^{-1}$ is also a generalized Lipschitz operator satisfying

$$\|(I + PC)^{-1}\|_{Lip} \le \|(I + PC)^{-1}(0)\|_Y + (1 - \|PC\|)^{-1}. \qquad (4.23)$$

Under the conditions mentioned above, it is easily seen from the system equations (4.17) that

$$e = (I + PC)^{-1}(W_1(u) - W_2(d)). \qquad (4.24)$$

Since $(I + PC)^{-1}$ is a nonlinear operator, unlike the linear case, we cannot separate the argument $W_1(u) - W_2(d)$ completely here in such a general setting. Observe, however, that $W_2 : D \to V$ is a bounded linear operator defined on the bounded closed subset D. By the so-called closed linear operator theorem, W_2 is a closed operator on D. Consequently, its generalized inner inverse

$W_2^+ : \mathcal{R}(W_2) \to D$ exists and is also a bounded linear operator (see Reference [8]). Here, a generalized inner inverse operator W_2^+ is defined to be one such that

$$W_2 W_2^+ W_2 = W_2$$

on the domain of W_2. This implies that

$$W_2 W_2^+ W_2(u) = W_2(u) \qquad \text{for all } u \in D,$$

and that on the range of W_2, we have $W_2 W_2^+ = I$ (the identity operator), namely,

$$W_2 W_2^+ (v) = v \qquad \text{for all } v \in \mathcal{R}(W_2).$$

Recall that we have assumed $\mathcal{R}(W_1) \subseteq \mathcal{R}(W_2)$. Hence, we also have

$$W_2 W_2^+ (v) = v \qquad \text{for all } v \in \mathcal{R}(W_1). \tag{4.25}$$

Moreover, the operator norm of this inner inverse operator satisfies

$$\|W_2^+\| := \sup_{\|v\|_Y = 1} \|W_2^+ v\|_Z \leq M_d, \tag{4.26}$$

since $W_2^+ v \in D$ for all $v \in \mathcal{R}(W_2)$. Consequently, it follows from (4.24) that

$$
\begin{aligned}
\|e_T\|_Y &= \|[(I + PC)^{-1}(W_1(u_T) - W_2(d_T))]_T\|_Y \\
&\leq \|(I + PC)^{-1} W_2\|_{Lip} \|[W_2^+ W_1(u_T)]_T - d_T\|_Z \\
&\leq \|(I + PC)^{-1} W_2\|_{Lip} \big(M_d \|W_1\| \|u_T\|_X + \|d_T\|_Z \big) \ (4.27)
\end{aligned}
$$

for all $T \in [0, \infty)$.

To simultaneously reject the disturbances and minimize the norm of the error response of the system, we may now formulate Problem 4.9 as follows:

$$\min_{C \in \mathcal{S}_0} \|(I + PC)^{-1} W_2\|_{Lip}. \tag{4.28}$$

This objective functional differs from the one for optimal tracking in that the operator W_2 has been taken into account here, which affects all the possible disturbance inputs.

If, however, for some special purposes on the system input signals u we require that the norm in the first term of the right-hand side of (4.27) remain less than or equal to a prescribed bound γ_0, namely,

$$\|(I + PC)^{-1}W_2\|_{Lip} \leq \gamma_0 \,,$$

then, since

$$\|W_2\|_{Lip} = \|(I + PC)(I + PC)^{-1}W_2\|_{Lip} \leq \|I + PC\|_{Lip}\gamma_0 \,,$$

by denoting $\gamma = \gamma_0^{-1}\|W_2\|_{Lip}$, we obtain

$$\gamma - \|I + PC\|_{Lip} \leq 0 \,.$$

In this case, the optimal disturbance rejection problem 4.9 may be posed as to minimize the second term in the right-hand side of (4.27) (due to the disturbances), subject to the constraint that the first term (due to the system input signals) remains less than or equal to a prescribed bound. This consideration leads to the following standard minimization formulated in terms of a Lagrangian functional:

$$\min_{\substack{C \in \mathcal{S}_0 \\ \lambda \in [0,\infty)}} L(C, \lambda) \,, \tag{4.29}$$

where λ is the Lagrangian multiplier and

$$L(C, \lambda) = \|(I + PC)^{-1}W_2\|_{Lip} + \lambda\{\gamma - \|I + PC\|_{Lip}\} \,,$$

with γ being a prescribed positive number as indicated above.

In the concern of the solvability of the problem, we observe that the objective functionals in both cases are continuous with respect to C (and λ), so that the optimization problems always have solutions under one of the two conditions stated in Theorem 4.5.

Recall that since the domain D of W_2 is only a bounded subset, the range of W_2 is in general a subset of the output space V. If $\mathcal{R}(W_2)$ is a bounded subset of V, then we can obtain one more condition under which the optimization problem is solvable. More precisely, we have the following:

4.10. Corollary. *Let $\mathcal{S}^b$ be a bounded subset in any finite-dimensional subspace of $\mathcal{S}$ and let*

$$\mathcal{S}_0 = \{C \in \mathcal{S}^b : \quad \|PC\| \le 1 - \epsilon\},$$

where $\epsilon > 0$ arbitrarily small. If the given plant operator P has the growth property that it maps an unbounded set to an unbounded set, then the optimization problem (4.28) or (4.29) has a solution in $\mathcal{S}_0$.

We remark that many nonlinear operators, including all the (nonzero) linear operators, have the growth property.

Proof. Recall that a closed bounded set in a finite-dimensional normed linear space is compact. Hence, based on the argument that any continuous functional assumes its infimum on a compact set in its domain, it is sufficient to show that the set $\mathcal{S}_0$ is bounded because it is obviously closed. Note that if P maps all unbounded sets to bounded sets, then $\mathcal{S}_0$ is not necessary bounded. However, due to the growth property of P, it can be verified that $\mathcal{S}_0$ must be bounded. To be more precise, suppose that $\mathcal{S}_0$ is not bounded, then there exists a sequence $\{C_n\}$ in it such that $\|C_n\|_{Lip} \to \infty$ as $n \to \infty$. Recall that W_2 is a bounded linear operator defined on a bounded set D; $\mathcal{R}(W_2)$ is a bounded set. Recall also that $\mathcal{D}(C) = \mathcal{R}(W_2)$, so that C_n are all defined on $\mathcal{R}(W_2)$; there is a sequence $\{e_n\}$ in $\mathcal{R}(W_2)$ such that $\|C_n(e_n)\|_Y \to \infty$ as $n \to \infty$. Hence, by the growth property of P, we have $\|PC_n(e_n)\|_Y \to \infty$ as $n \to \infty$, so that

$$\|PC_n\|_{Lip} \ge \|PC_n(0)\|_Y + \frac{\|PC_n(e_n)\|_Y}{M} \to \infty \quad \text{as } n \to \infty,$$

where M is the radius of a ball containing the bounded subset $\mathcal{R}(W_2)$ in V. This contradicts the condition that $\|PC_n\|_{Lip} \le 1 - \epsilon$ for all $C_n \in \mathcal{S}_0$. $\qquad\square$

§ 4. Robust Stabilization

Consider a general MIMO nonlinear closed-loop system defined in an extended linear space setting as shown in Figure 4.8.

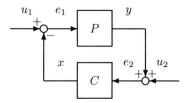

Figure 4.8.   A nonlinear feedback system.

The system equations are given by

$$\begin{cases} e_1 = u_1 - C(e_2) \\ e_2 = u_2 + P(e_1), \end{cases} \tag{4.30}$$

where u_1 and u_2 are system inputs considered as reference signals, e_1 and e_2 are corresponding error signals caused by feedback and feedforward, respectively, and P and C are nonlinear plant and compensator operators, respectively.

Let the spaces X^e (to which u_1, e_1, and x belong) and Y^e (to which u_2, e_2, and y belong) be two extended linear spaces of real vector-valued measurable functions defined as usual on the time domain $[0, \infty)$. Again, using the notation $\mathcal{D}$ and $\mathcal{R}$ for the domain and range, respectively, of operators, we assume for consistency that $\mathcal{D}(P), \mathcal{R}(C) \subseteq X$ and $\mathcal{D}(C), \mathcal{R}(P) \subseteq Y$. Also, let the input and output spaces be U and V, respectively, with

$$U \subseteq X^e \text{ where } U = \left\{ u : \quad \|u_T\|_X < \infty \text{ for all } T \in [0, \infty) \right\} \tag{4.31}$$

and

$$V \subseteq Y^e \text{ where } V = \left\{ v : \quad \|v_T\|_Y < \infty \text{ for all } T \in [0, \infty) \right\}. \tag{4.32}$$

We require that $u_1, e_1, x \in U$ and $u_2, e_2, y \in V$ for the input-output stability, where U is the input space for P but is the output space for C, and V is the input space for C but the output space for P.

In addition, we assume that the given nonlinear plant operator P has a nonlinear perturbation with respect to the nonlinear nominal plant P_0, namely,

$$P = P_0 + \Delta P, \qquad (4.33)$$

in which both P_0 and the perturbation ΔP are assumed to be generalized Lipschitz operators, and by a suitable rescaling of the norm of P (and the whole system, if necessary), we may require

$$\|\Delta P\|_{Lip} \le \epsilon_0 < 1 \qquad (4.34)$$

for some constant ϵ_0.

As to the compensator C to be designed, we will require it be a generalized Lipschitz operator as usual. The problem under consideration here is called the *robust stabilization* problem and is stated as follows:

4.11. Problem. *Given P_0, U, and V as described above, design a compensator C belonging to a certain admissible class S so as to stabilize the overall closed-loop system while maximizing the possible range of the perturbation ΔP.*

As mentioned before, we will only give a brief discussion on this problem here in this section since all the ideas and techniques are similar to that in the discussions of the optimal tracking problem.

Before we formulate the problem precisely in a mathematical manner, we establish the following lemma, which characterizes stabilizability of the system. In doing so, we need to use the sup-norm for the composite operator of P and C defined by (3.18), namely,

$$\|P(* - C(\cdot))\| := \|P(-C(0))\|_Y + \sup_{\substack{T \in [0,\infty) \\ u \ne 0}} \sup_{\substack{u \in U \\ [\tilde{e}_1]_T \ne [\tilde{e}_2]_T}}$$

$$\frac{\|[P(u_T - [C([\tilde{e}_1]_T)]_T)]_T - [P(u_T - [C([\tilde{e}_2]_T)]_T)]_T\|_Y}{\|u_T\|_X + \|[\tilde{e}_1]_T - [\tilde{e}_2]_T\|_Y}. \quad (4.35)$$

As mentioned before, this is a mixed two-variable operator norm, and for each fixed $u \in U$ or fixed $\tilde{e}_1, \tilde{e}_2 \in V$, it is a generalized Lipschitz norm.

4.12. Lemma. *Let P_0 be a given nominal plant that is assumed to be a generalized Lipschitz operator. Then, the family S defined below consists of compensators C that stabilize the system:*

$$S = \left\{ C \in Lip(V,U) : \quad \|P(* - C(\cdot))\| < 1 \right\}. \qquad (4.36)$$

This lemma can be easily verified by modifying the proof of Theorem 3.14.

Now, observe that from (4.33) we have

$$P(u - C(e)) = P_0(u - C(e)) + \Delta P(u - C(e)) \qquad (4.37)$$

and from (4.36) we have

$$\|P(* - C(\cdot))\| < 1. \qquad (4.38)$$

Since the problem is to maximize the possible range of the plant perturbation ΔP under the condition of stability, in view of (4.37) we may minimize the operator norm $\|P_0(* - C(\cdot))\|$ under the constraint $\|P_0(* - C(\cdot))\| < 1 - \epsilon_0$ which, together with (4.34) and (4.38), guarantees the stability condition:

$$\|P(* - C(\cdot))\| \leq \|P_0(* - C(\cdot))\| + \|\Delta P(* - C(\cdot))\|$$
$$\leq \|P_0(* - C(\cdot))\| + \epsilon_0$$
$$< 1.$$

Hence, the problem can be posed as follows.

4.13. Problem.

$$\min_{C \in S} \|P_0(* - C(\cdot))\|,$$

where S is defined by

$$S = \left\{ C \in Lip(V,U) : \quad \|P_0(* - C(\cdot))\| < 1 - \epsilon_0 \right\}.$$

We remark that since the objective and the constrained functionals are the same, this problem is equivalent to the following:

$$\min_{C \in Lip(V,U)} \|P_0(* - C(\cdot))\|, \qquad (4.39)$$

where the infimum of the objective functional is understood to be less than $1 - \epsilon_0$, and otherwise we conclude that the problem has no solution. We will, however, only consider the problem stated in 4.13 in the following.

Similar to Theorem 4.5, we have the following result:

4.14. Theorem. *The objective functional*

$$f(C) := \|P_0(* - C(\cdot))\|$$

is continuous with respect to C. *Consequently, Problem* 4.13 *is solvable under one of the following conditions:*

(i) *Let* $Lip^c(V, U)$ *be a compact subset of* $Lip(V, U)$ *and let*

$$S_0 = \left\{ C \in Lip^c(V, U) : \quad \|P_0(* - C(\cdot))\| < 1 - \epsilon_0 \right\}.$$

    Then, Problem 4.13 *has a solution in* S_0.

(ii) *Let* $Lip^b(V, U)$ *be a bounded set in any finite-dimensional subspace of* $Lip(V, U)$ *and let*

$$S_0 = \left\{ C \in Lip^b(V, U) : \quad \|P_0(* - C(\cdot))\| \le 1 - \epsilon \right\},$$

where $\epsilon > \epsilon_0$. *Then Problem* 4.13 *has a solution in* S_0.

Also, similar to Corollary 4.6, the following result is true, as may be expected.

4.15. Corollary. *Let*

$$\delta = \inf_{C \in S_k} \|P_0(* - C(\cdot))\|$$

and

$$\delta_k = \inf_{C_k \in S_k} \|P_0(* - C_k(\cdot))\|,$$

where S_k *is defined either in* (i) *or* (ii) *in Theorem* 4.14 *in a* k-*dimensional subspace of* S. *Then, we have* $\delta_k \to \delta$ *as* $k \to \infty$.

We close this chapter with a remark. All the optimal design problems described in this chapter have the same underlying framework, which is the nonlinear generalized Lipschitz operators theory established in an extended linear space setting in the time domain. Since the notion of standard transfer functions does not apply to nonlinear control systems, formulating such design problems in the time domain seems to be natural. The advantages of this generalized Lipschitz operators approach, as mentioned before, include that the overall nonlinear feedback system can be well posed physically as well as mathematically, in the sense that all nonlinear operators involved (including the inherent nonlinear inverse operator but excluding perhaps the given plant operator) are causal and belong to the same (infinite-dimensional) Banach space of generalized Lipschitz operators, and all internal signals are uniquely determined by their corresponding input signals. These elegant features, as is well known, are difficult to achieve from any other approaches in general. Due to the generality of the nonlinear framework under consideration, it has not been possible to touch upon many detailed issues at the technical level in the above discussions of feedback systems design. However, the theoretical analysis given throughout the chapter has provided many significant insights about the global structure of a general nonlinear feedback control system, which can be used as useful principles and references for other purposes of optimal design in applications.

Exercises

4.1. Give examples to show that even if all the operators, W_1, W_2, C, P, shown in Figure 4.5 are bounded in certain operator norms, the overall closed-loop system may still be input-output unstable.

4.2. Show that the Lipschitz operator $(I + PC)^{-1}$, given in (4.13), and the functional $\|(I + PC)^{-1}\|_{Lip}$ both depend on C continuously.

4.3. Show that a continuous functional assumes its maximum and minimum on any compact subset in its domain. Give examples to show that this is not true in general if the subset is only bounded and closed but not compact.

4.4. Show that the polynomic operators or nonlinear Volterra operators of finite degree constitute a dense set in the Banach space $\mathcal{S}$ of Lipschitz operators defined in (4.10).

4.5. Give examples of generalized Lipschitz operators, the total of which consists of a compact set in the infinite-dimensional space $\mathcal{S}$ defined by (4.10).

4.6. Show that a compact metric space is separable, in the sense that it has a countable basis.

4.7. Are Problems 4.4 and 4.8 equivalent for the SISO linear setting? Prove or disprove your claim.

4.8. Verify that the functional $L(C, \lambda)$ given in (4.29) is continuous with respect to C (and λ).

4.9. Show that a closed bounded set in a finite-dimensional normed linear space is compact. Give examples to show that this is not true in general if the space is infinite-dimensional.

4.10. Consider the single-link robot arm system discussed in Section 4.1, with the inexact modeling part being formulated as a disturbance, as shown in Figure 4.4. Design a nonlinear controller to replace the PD controller therein, such that the overall feedback control system is stable and can reject the disturbance optimally, while the system output is tracking the reference input.

References

[1] G. Chen and R.J.P. de Figueiredo, "Optimal nonlinear feedback system design for a general tracking problem," in *Mathematical Theory of Networks and Systems*, (M.A. Kaashoek *et al*, eds.), pp. 429–436, Birkhäuser, Boston, 1989.

[2] G. Chen and R.J.P. de Figueiredo, "On robust stabilization of nonlinear control systems," *Sys. Contr. Letts.*, **12** (1989), 373–379.

[3] C.K. Chui and G. Chen, *Signal Processing and Systems Theory: Selected Topics*, Springer-Verlag, New York, 1992.

[4] R.J.P. de Figueiredo and G. Chen, "Optimal disturbance rejection for nonlinear control systems," *IEEE Trans. Auto. Contr.*, **12** (1989), 1242–1248.

[5] C.A. Desoer and M. Vidyasagar, *Feedback Systems: Input-Output Properties*, Academic Press, New York, 1975.

[6] V. Dolezal, "Sensitivity and robust stability of general input-output systems," *IEEE Trans. Auto. Contr.*, **36** (1991), 539–550.

[7] R.H. Martin, *Nonlinear Operators and Differential Equations in Banach Spaces*, Wiley, New York, 1976.

[8] M.Z. Nashed, "On generalized inverses and operator ranges," *Functional Analysis and Approximation*, (P.C. Butzer *et al*, eds.), pp.85–96, Birkhäuser, Boston, 1981.

Chapter 5

Coprime Factorizations of Nonlinear Mappings for Control Systems

In this chapter, a special topic of coprime factorizations for nonlinear mappings that describe control systems will be studied. Coprime factorization techniques for linear mappings that describe control systems have been well developed, which, as is well known, is very important in the study of robust stabilization of linear control systems. To formulate a nonlinear analogue of the coprime factorization, a commonly used input-output stability criterion is adopted here, and the previously developed generalized Lipschitz operators theory and extended linear spaces framework will be applied. After describing all necessary concepts and terminologies in Section 1, such as right and left factorizations as well as coprimeness for nonlinear mappings that describe control systems, right and left coprime factorizations for the standard feedback configuration will be formulated in Section 2. Then, in Section 3, a simple necessary and sufficient condition on the existence of a right coprime factorization will be developed, and a simple sufficiency result along with its associated constructive scheme for obtaining right (and left) coprime factorizations of nonlinear feedback control systems will be derived. An illustrative example will be given in Section 4. Another formulation for the left (and right) coprime factorization for a class of general nonlinear control systems described by vector-valued ordinary differential equations will then be discussed in Section 5, where a concrete algorithm for constructing such a coprime factorization will also be demonstrated with an example. Finally, in Section 6, a nonlinear feedback configuration under a formulation of the left coprime factorization will be discussed briefly.

§ 1. Preliminaries

In order to introduce the notion of coprime factorizations for non-linear mappings that describe control systems, we first have to clarify some basic notation and definitions that are needed throughout this chapter.

We first recall the concept of input-output stability that is of central importance to us in this chapter. In the following, we will use the notation $\mathcal{D}$ and $\mathcal{R}$, respectively, for the domain and range of the corresponding nonlinear operators.

5.1. Definitions. Let S_i^e and S_o^e be two prescribed extended linear spaces associated with two Banach spaces S_i and S_o of real vector-valued measurable functions defined on the time domain $[0, \infty)$, respectively. S_i and S_o will be called the *input space* and *output space* respectively in the following. Let $T : \mathcal{D}(T) \to \mathcal{R}(T)$ be a non-linear operator with its domain $\mathcal{D}(T) \subseteq S_i^e$ and range $\mathcal{R}(T) \subseteq S_o^e$. T is said to be *input-output stable*, or simple *stable* throughout this chapter, if T maps all input functions from S_i into the output space S_o. In notation, T is stable if and only if $T(S_i) \subseteq S_o$. Otherwise, namely, if T maps some input functions from $S_i \cap \mathcal{D}(T)$ to the set $\mathcal{R}(T) \backslash S_o$ (if not empty), it is said to be *unstable*.

To facilitate our discussions below, we will always assume that $S_o \subset \mathcal{R}(T)$, so that $S_o \subset \mathcal{R}(T) \subseteq S_o^e$, unless otherwise indicated. Similarly, we assume that $S_i \subseteq \mathcal{D}(T) \subseteq S_i^e$. The reason for excluding the case $S_o = \mathcal{R}(T)$ here is to consider the possibility of instability for the operator T in a general discussion.

Based on these definitions, we have the following concept of stabilizability.

5.2. Definitions. An unstable nonlinear operator $T : \mathcal{D}(T) \to \mathcal{R}(T)$ is said to be *stabilizable* if there exists an operator $F : \mathcal{D}(T) \to \mathcal{D}(T)$ such that the composite operator $TF = T \circ F$ is stable from S_i to S_o.

A simple example of unstabilizable nonlinear operators is the "constant" operator, which maps all inputs from its input space to a (fixed) function located outside the output space.

Note that there always exists a nonlinear operator $F : \mathcal{R}(T) \to \mathcal{R}(T)$ such that the composite operator FT is stable from S_i to S_o, provided that $S_o \neq \emptyset$, since we can define F to be the nonlinear operator that maps every function to a (fixed) function located inside S_o.

To introduce the next concept, we first remark that there are several different ways of defining the operator norm for a nonlinear operator T on its domain $\mathcal{D}(T)$. Whenever an operator norm $\| \cdot \|_o$ for T has been defined and it turns out to be finite, $\|T\|_o < \infty$, T is said to be of *finite gain* (bounded in an operator norm). If moreover the operator T is stable, then it is said to be *finite-gain stable*. We will only consider finite-gain stability in this chapter. It is important to note that a finite-gain operator may not be (input-output) stable since it may map some input function from the input space to somewhere in its range which is outside the output space.

We now introduce the concepts of right and left factorizations of causal nonlinear operators.

5.3. Definitions. Let $T : \mathcal{D}(T) \to \mathcal{R}(T)$ be a causal and stabilizable (but need not be stable) nonlinear operator defined as above with the prescribed input-output spaces S_i and S_o. T is said to have a *right factorization* over the space of finite-gain stable and causal operators if there exist two finite-gain stable and causal operators $D : \mathcal{D}(T) \to \mathcal{D}(T)$ and $N : \mathcal{D}(T) \to \mathcal{R}(T)$, where the input space for both D and N is the original input space S_i associated to T, the output space for D is also S_i, and the output space for N is the original output space S_o, such that $D^{-1} : \mathcal{D}(T) \to \mathcal{D}(T)$ exists, is causal (but not necessarily stable) on (a subset of) $\mathcal{D}(T)$ containing S_i, and such that $T = ND^{-1}$ on $\mathcal{D}(T)$. Here, the input and output spaces for D^{-1} are both S_i. A *left factorization* for T over the space of finite-gain stable and causal operators is defined in a similar manner such that $T = D^{-1}N$ on $\mathcal{D}(T)$, with suitably defined domains, ranges, and input and output spaces for the operators involved.

Here, it is necessary to clarify the reasoning for our choices of domains, ranges, and input-output spaces of various operators shown in the above definitions. It is clear that if an operator T has a right factorization on its domain over the space of finite-gain stable and causal operators, $T = ND^{-1}$ on $\mathcal{D}(T)$, then we have $TD = N$ on $\mathcal{D}(T)$, which implies that the composite operator TD must be stable since N is stable. Consequently, T has to be stabilizable and moreover T and D must have the same input space and T and N must have the same output space. On the other hand, it follows from the right factorization $T = ND^{-1}$ that T and D^{-1} must have the same input space, and for the overall stability and well-composition of ND^{-1} the output space of D^{-1} may be chosen to be the input space of N. In addition, we have ensured that all the domains and ranges of the operators D, N, and D^{-1} satisfy our convention

$$S_i \subseteq \mathcal{D}(T) \subseteq S_i^e \qquad \text{and} \qquad S_o \subset \mathcal{R}(T) \subseteq S_o^e,$$

as can be seen from the above definitions.

We finally have the following definitions.

5.4. Definitions. Let $T : \mathcal{D}(T) \to \mathcal{R}(T)$ be a causal and stabilizable (but need not be stable) nonlinear operator defined as above. T is said to have a *right coprime factorization* on $\mathcal{D}(T)$ over the space of finite-gain stable and causal operators if it has a right factorization $T = ND^{-1}$ on $\mathcal{D}(T)$ over the space of finite-gain stable and causal operators and moreover there exist two operators $A : \mathcal{R}(N) \to \mathcal{D}(T)$ and $B : \mathcal{R}(D) \to \mathcal{D}(T)$, where the output spaces for A and B are both the original input space S_i associated to T, the input space for A is the original output space S_o associated to T, and the input space for B is S_i, such that A and B are both finite-gain stable and causal and such that for the identity operator $I : \mathcal{D}(T) \to \mathcal{D}(T)$, we have the Bezout identity:

$$AN + BD = I. \qquad (5.1)$$

Note that the operator $B = (I - AN)D^{-1}$ (obtained from (5.1)) is stable from $\mathcal{D}(T)$ to $\mathcal{D}(T)$. Moreover, when a nonlinear feedback system is considered, as will be seen in the next section, we will also require the operator B be such that $B^{-1} : \mathcal{D}(T) \to \mathcal{D}(T)$ exists and is finite-gain stable and causal on $\mathcal{D}(T)$. This viewpoint will be illustrated later.

We now summarize the relationships among all the above-mentioned domains, ranges, and input-output spaces of the operators involved in Table 5.1, in which we always assume that

$$S_i \subseteq \mathcal{D}(T) \subseteq S_i^e \qquad \text{and} \qquad S_o \subset \mathcal{R}(T) \subseteq S_o^e.$$

For the left coprime factorization, we can easily derive similar conditions on the domains, ranges, and input-output spaces of the nonlinear operators involved for the validity of the expressions

$$T = D^{-1}N \quad \text{and} \quad NA + DB = I. \tag{5.2}$$

Table 5.1.

Operators	Domains	Input spaces	Output spaces	Ranges
T	$\mathcal{D}(T)$	S_i	S_o	$\mathcal{R}(T)$
N	$\mathcal{D}(T)$	S_i	S_o	$\mathcal{R}(T)$
D	$\mathcal{D}(T)$	S_i	S_i	$\mathcal{D}(T)$
D^{-1}	$\mathcal{D}(T)$	S_i	S_i	$\mathcal{D}(T)$
A	$\mathcal{R}(N) = \mathcal{R}(T)$	S_o	S_i	$\mathcal{D}(T)$
B	$\mathcal{R}(D) = \mathcal{D}(T)$	S_i	S_i	$\mathcal{D}(T)$
B^{-1}	$\mathcal{D}(T)$	S_i	S_i	$\mathcal{D}(T)$

§ 2. Right Coprime Factorization of Nonlinear Feedback Systems

In this section, we consider the following nonlinear feedback system shown in Figure 5.1, where X^e, Y^e, and Z^e are extended linear spaces associated with three given Banach spaces X, Y, and Z, respectively, of real vector-valued measurable functions defined on the time domain $[0, \infty)$; P is a given nonlinear plant operator that is assumed to be causal and stabilizable but not necessarily stable; and A and B are two nonlinear operators to be designed according to some criteria to be described later, in which it will be required that $Q := B^{-1}$ exists on the subset S of X^e defined by

$$S = \{e \in X^e : \quad e = u - x, \ u \in X^e, x \in A(Y^e)\}, \tag{5.3}$$

where

$$A(Y^e) = \{x \in X^e : \quad x = A(y), \ y \in Y^e\}$$

for a pre-designed operator A.

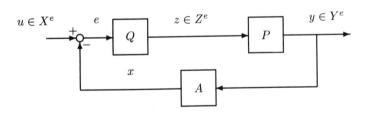

Figure 5.1.   A nonlinear feedback system.

Here, the reason for setting $Q = B^{-1}$ is the following: Consider the input and error-output relationship

$$z = Q(u - A(y)) = Q(u - AP(z)).$$

If P has a right factorization $P = ND^{-1}$ and if Q is invertible on their domains, then we arrive at

$$\begin{aligned} z &= [Q^{-1} + AP]^{-1}(u) \\ &= [Q^{-1} + AND^{-1}]^{-1}(u) \\ &= D[Q^{-1}D + AN]^{-1}(u). \end{aligned}$$

Consequently, by letting $B := Q^{-1}$ and $M := BD + AN$, we obtain

$$z = DM^{-1}(u),$$

so that the feedback system is stable, in the concern of input/error-output relation, whenever M is stable and has a stable inverse M^{-1}. Hence, it is particularly convenient to set $Q = B^{-1}$ and to require operators A, B, and B^{-1} be stable. Moreover, since both M and M^{-1} exist and are stable, without loss of generality we may simply reduce M to be the identity operator I and impose the condition of the validity of the Bezout identity

$$AN + BD = I$$

on the domain of the plant operator P, which has the right factorization $P = ND^{-1}$ on its domain. This also gives a motivation for the consideration of the Bezout identity for the feedback system shown in Figure 5.1.

In order for this feedback system to be well posed in the sense of stability, causality, and uniqueness of internal signals, further assumptions, conditions, and requirements have to be clarified precisely.

First, we only consider the generalized Lipschitz norm for the system operators defined on extended linear spaces in the following. Throughout this chapter, by a Lipschitz operator we will always mean one defined in the generalized sense. The advantages of using the Lipschitz operator norm have been emphasized previously, which will be demonstrated once again in a different situation in this chapter.

In the following, for the family of stable nonlinear Lipschitz operators T mapping from its domain $\mathcal{D}(T)$ (which coincides with its input space X) into its range $\mathcal{R}(T)$ (which contains its output space Y), we will use the following notation, which is slightly different from what has been used before:

$$Lip(X, Y) = \{T : \mathcal{D}(T) = X, \ Y \subset \mathcal{R}(T) \subseteq Y^e,$$
$$T(X) \subseteq Y, \ \|T\|_{Lip} < \infty \text{ on } X\}. \quad (5.4)$$

However, for a family of general Lipschitz operators without taking the input-output spaces into account, such as the operator D^{-1}

encountered later, we will not use this notation in this chapter. In other words, $Lip(X,Y)$ is a family of finite-gain stable and causal nonlinear Lipschitz operators with the input-output spaces X and Y. Similar to Theorem 2.21, it can be easily proved that $Lip(X,Y)$ is an infinite-dimensional Banach space. Hence, a family of nonlinear Lipschitz operators of this kind is actually a very large class, which, as mentioned before, contains all bounded linear operators defined on X with the ranges in Y^e.

Now let us return to the feedback control system shown in Figure 5.1. Suppose that the extended linear spaces X^e, Y^e, and Z^e, together with their subspaces X, Y, Z and the set $S \subseteq X^e$, are all given. Assume, furthermore, that X and Y are input-output spaces and the nonlinear plant operator $P : \mathcal{D}(P) \to \mathcal{R}(P)$ with $\mathcal{D}(P) = Z$ and $\mathcal{R}(P) \subseteq Y^e$ is given. P is not necessarily stable nor Lipschitz (not even of norm-bounded), but must be stabilizable.

It will be seen below that all nonlinear operators (except probably the given plant P) involved in the feedback configuration, including their compositions and inverses, are Lipschitz and hence are of finite gain and causal. Consequently, the overall feedback system is well posed in the sense that all nonlinear operators involved (except probably the plant P) belong to the same family of finite-gain and causal operators and that all internal signals are uniquely determined by their corresponding inputs.

We have not yet mentioned the stability. As emphasized before, the norm-boundedness of an operator does not necessarily imply the input-output stability, since the stability actually depends on the prescribed input-output spaces. In view of Table 5.1, with $T = P$ therein, we see that we should set $\mathcal{D}(P) = Z, \mathcal{R}(P) = Y^e$, and let the input-output spaces for P be $S_i = Z$ and $S_o = Y$, respectively. In order for the given plant operator P to have a right factorization $P = ND^{-1}$, an operator $D \in Lip(Z,Z)$, with $\mathcal{R}(D) = Z^e$, has to exist such that D^{-1} exists and is causal on Z and such that the composite operator $PD \in Lip(Z,Y)$. We will assume this existence here and then consider sufficiency conditions for construction of solutions later. Hence, we set

$$\mathcal{S} = \{\, D \in Lip(Z,Z): \quad D^{-1} \text{ exists and is causal on } Z$$
$$\text{such that } PD \in Lip(Z,Y) \,\}, \quad (5.5)$$

and assume that $\mathcal{S} \neq \emptyset$. Otherwise, the given plant P has no right factorization. If the given plant P is Lipschitz and stable, namely, if $P \in Lip(Z, Y)$, then clearly the identity operator $I \in \mathcal{S}$. Suppose that P is unstable (but stabilizable) and that an operator $D \in \mathcal{S}$ has been chosen, under certain specific criteria (for the easy construction of a nontrivial right factorization for P, for instance, which is not important at this moment). Then, we see that D is finite-gain stable and causal since it maps all inputs from its input space Z into the prescribed output space Z, but D^{-1} may not. What we know about the operator D^{-1} is that it exists and is causal on the Banach subspace Z such that $PD : Z \to Y^e$ is causal and finite-gain stable (with the input-output spaces Z and Y). Set

$$N = PD. \tag{5.6}$$

Then, $N : Z \to Y^e$ is causal and finite-gain stable (with the input-output spaces Z and Y). We have therefore obtained a well-defined right factorization for the nonlinear plant operator P, namely,

$$P = ND^{-1}, \tag{5.7}$$

which is valid on $\mathcal{D}(P) = Z$. Obviously, P is stable (from Z to Y) if D^{-1} is stable (from Z to Z), where D^{-1} is stable if $D^{-1}(Z) \subseteq Z$, and P is causal and finite-gain stable (from Z to Y) if so is D^{-1}, namely, if $D^{-1} \in Lip(Z, Z)$. The converse is not true in general.

Our next goal is to find two operators A and B to satisfy the Bezout identity and hence to stabilize the overall feedback system.

In order to pose the problem precisely and to ensure that the overall feedback system be well posed so as a right coprime factorization can be achieved, as well as to make the following discussions easier, we may simply let

$$X = Y = Z \quad \text{and} \quad X^e = Y^e = Z^e.$$

Moreover, to handle the difficulty in the definition of the set $\mathcal{S}$ in (5.3), we may simply let $\mathcal{S} = X^e$. Consequently, the set $\mathcal{S}$ defined in (5.5) becomes

$$\mathcal{S} = \{ \, D \in Lip(X, X) : \quad D^{-1} \text{ exists and is causal on } X$$
$$\text{such that } PD \in Lip(X, X) \, \}. \tag{5.8}$$

In summary, for the consistency of the overall feedback configuration shown in Figure 5.1 and to facilitate our discussions, we
finally specify the conditions on the domains, ranges, and input-
output spaces of the nonlinear operators involved and consider the
closed-loop configuration shown in Figure 5.2.

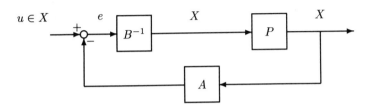

Figure 5.2.   A nonlinear feedback system.

We are now in a position to state the right coprime factorization problem very precisely. The left coprime factorization problem
can be formulated in a similar manner, which will not be discussed
in detail below.

5.5. Problem.   *Given an extended linear space X^e associated
with a Banach space X of real vector-valued measurable functions
defined on the time domain $[0, \infty)$. Given also a nonlinear plant
operator $P : X \to X^e$, which need not be stable nor Lipschitz (but is
necessarily stabilizable). Moreover, assume that the set S defined
in (5.8) be given and nonempty, and let an operator $D \in S$ be
chosen and fixed. Set $N = PD$, so that P has a right factorization
$P = ND^{-1}$. Find a pair of operators $A, B \in Lip(X, X)$ such that*

$$AN + BD = I, \qquad (5.9)$$

*where $I : X \to X$ is the identity operator, and such that $B^{-1} \in
Lip(X, X)$.*

We will give simple yet concrete sufficiency conditions for construction of solutions of the problem and in the meantime show
a constructive scheme for solving the problem in the next section.
For convenience, we close this section with reference to Table 5.2,
which shows the domains, ranges, and input-output spaces of all
the operators involved in Figure 5.2.

Table 5.2.

Operators	Domains	Input spaces	Output spaces	Ranges
P	X	X	X	X^e
N	X	X	X	X^e
D	X	X	X	X^e
D^{-1}	X^e	X	X	X^e
A	X^e	X	X	X
B	X	X	X	X
B^{-1}	X	X	X	X

§ 3. Necessary and Sufficient Condition and Construction of Coprime Factorizations

In this section, we first derive a simple necessary and sufficient condition for the existence of a right coprime factorization of a nonlinear mapping that describes a control system. This condition indeed characterizes solutions of Problem 5.5 posed in the last section.

The necessary and sufficient condition on the existence of a right coprime factorization of the nonlinear feedback system shown in Figure 5.1 is stated in the following. It is interesting to observe that the condition derived here is actually independent of the stability concept, in the sense that the theorem statement and its proof need not involve any stability, as can be easily verified by removing the stability throughout. We present the stability requirement in this result because the stability is important for control systems under consideration (and hence has been imposed in our definitions given above).

5.6. Theorem. The feedback control system shown in Figure 5.1 possesses a right coprime factorization if and only if the composite operator $(I + APB^{-1}) : \mathcal{R}(B) \to X^e$ is one-to-one and onto, its inverse is causal, and all the operators A, B, D, N, B^{-1}, and $(I + APB^{-1})^{-1}$ are causally stable.

Here, we should recall that a one-to-one and onto operator has a (two-sided) inverse.

Proof. First, observe that if the system has a right coprime factorization, then there exist two causal operators $N : X \to X^e$ and $D : X \to X^e$ with a causal inverse $D^{-1} : X^e \to X$ such that $ND^{-1} = P$ and $AN + BD = I$. Since $B : X \to X$ is one-to-one, with domain X, and is onto $\mathcal{R}(B)$, we see that $\mathcal{R}(B^{-1}) = X \subset \mathcal{D}(D^{-1})$. On the other hand, observe that $\mathcal{R}(D) = X^e$. It hence follows that

$$I + APB^{-1} = I + AND^{-1}B^{-1}$$
$$= [BD + AN]D^{-1}B^{-1} = D^{-1}B^{-1},$$

which implies that the operator $(I + APB^{-1}) : \mathcal{R}(B) \to X$ is causal, one-to-one and onto, and hence is causally invertible, with the inverse equal to $BD : X \to \mathcal{R}(B)$.

Conversely, if the operator $(I + APB^{-1}) : \mathcal{R}(B) \to X$ is causally invertible, then we can define

$$D = B^{-1}(I + APB^{-1})^{-1} : \quad X \to X^e \, .$$

Since $(I + APB^{-1})^{-1} : X \to \mathcal{R}(B)$ is onto, we have $\mathcal{R}((I + APB^{-1})^{-1}) = \mathcal{R}(B) = \mathcal{D}(B^{-1})$. On the other hand, observe that $\mathcal{R}(B^{-1}) = X$. Hence, the operator D so defined is causal, one-to-one and onto. It follows that

$$D = [(I + APB^{-1})B]^{-1} = (B + AP)^{-1} \, .$$

Consequently, $D^{-1} = B + AP : X^e \to X$ exists and is causal, so that if we define $N = PD$, then N is causal, and moreover we obtain both $ND^{-1} = PDD^{-1} = P$ and

$$AN + BD = (AND^{-1} + B)D$$
$$= (AP + B)D = D^{-1}D = I \, .$$

This, along with the stability conditions imposing on the related operators, completes the proof of the theorem. $\qquad\Box$

The necessary and sufficient condition for the existence of a right coprime factorization of a nonlinear mapping that describes a control system is quite simple and elegant. However, it does not seem to be easy to use to construct a coprime factorization. Hence, we need to search for some practical methods for the construction.

In the following, we show some sufficiency conditions for constructing solutions for Problem 5.5 posed in the previous section. We will describe a simple scheme to find pairs of the two nonlinear operators A and B. In the algorithm derived below, although it is impossible to find all solutions in general, we can find a very large set of solutions. In applications, however, one solution is usually satisfactory for many purposes.

To describe the sufficiency conditions and constructive scheme, we only discuss in detail right coprime factorizations, since left coprime factorizations can be done in the same manner.

Given an extended linear space X^e together with its associated Banach space X (the input-output spaces for the feedback system) and given a nonlinear plant operator $P : X \to X^e$ (with X as input-output spaces), which is not necessarily stable but has to be stabilizable (from X to X). Moreover, P is not necessarily Lipschitz nor norm-bounded (in any operator norm). Recall, however, that we have assumed that the set S defined by (5.8) is nonempty so that the coprime factorization problem is solvable and hence a construction of the coprime factorization is possible.

We first choose and fix an arbitrary operator D in the set S defined in (5.8), namely,

$$S = \{ D \in Lip(X,X) : \quad D^{-1} \text{ exists and is causal on } X$$
$$\text{such that } PD \in Lip(X,X) \}, \text{(5.10)}$$

and then set $N = PD$. Since $N = PD \in Lip(X,X)$, which is causal and finite-gain stable (from X to X), we have already obtained a right factorization $P = ND^{-1}$ as desired, which is valid on X.

Next, we can construct the operator B using the formula

$$B = (I - AN)D^{-1}, \tag{5.11}$$

in which the operator A is chosen from the following set:

$$A_1 = \{A \in Lip(X,X) : \quad (I - AN)D^{-1} \in Lip(X,X)\}. \tag{5.12}$$

Recall that D^{-1} is unstable (from X to X) if the given plant P is so. Hence, in this case any operator $A \in A_1$ will stabilize the unstable operator D^{-1}. Quite often, there will be a cancellation of unstable modes between the two operators D^{-1} and $(I - AN)$, so that the composite operator $(I - AN)D^{-1}$ turns out to be stable. This is known for the linear systems and can also be seen from the nonlinear example shown in Section 4 below. In this case, the set $A_1 \neq \emptyset$. Otherwise, a stable operator B cannot be found by using Formula (5.11), and hence the Bezout identity (5.9) (which is equivalent to (5.11) provided that D^{-1} exists) will not hold, so that the coprime factorization problem has no solution.

As to the operator B^{-1}, it follows from (5.11) that

$$B^{-1} = D(I - NA)^{-1}, \tag{5.13}$$

provided that $(I - NA)^{-1}$ exists. Since D is stable (from X to X), in order for the operator B^{-1} to be stable in the same sense, it is necessary that the inverse operator $(I - AN)^{-1}$ be stable, also from X to X, unless there is a cancellation of unstable modes between the two operators $(I - AN)^{-1}$ and D. In general, if there is no cancellation of unstable modes, then in order to guarantee the stability of the composite operator $D(I - AN)^{-1}$, we can always pick an operator A in the following set:

$$\mathcal{A}_2 = \{A \in Lip(X, X): \quad \|AN\| < 1\}, \qquad (5.14)$$

where $\|\cdot\|$ is the semi-norm for the Lipschitz operators as defined in (2.17). Any $A \in \mathcal{A}_2$ will guarantee $(I - AN)^{-1} \in Lip(X, X)$ based on Theorem 2.22. Moreover, since $\|AN\| \leq \|N\| \|A\|$, all operators A satisfying $\|A\| < \|N\|^{-1}$ belong to the set $\mathcal{A}_2$, where $N \neq 0$ unless $P = 0$. Indeed, it follows from Theorem 2.21 that $\mathcal{A}_2$ is an open set (containing an open ball) in an infinite-dimensional Banach space of operators and is hence a very large set. Of course, the condition $A \in \mathcal{A}_2$ is only a sufficient one that may not be necessary, especially in the case that there is a cancellation of unstable modes between two operators as pointed out above and as can be seen from the example shown in Section 4 below.

We now summarize the above-described sufficient conditions for obtaining right coprime factorizations for the nonlinear feedback system shown in Figure 5.2, in which, again, the nonlinear plant operator P is not necessarily Lipschitz nor stable (but has to be stabilizable) on its domain X. As long as the sets $\mathcal{S} \neq \emptyset$, $\mathcal{A}_1 \neq \emptyset$ and $\mathcal{A}_2 \neq \emptyset$, we have the following sufficiency conditions, which provide as well a construction scheme:

5.7. Algorithm.
(1) Choose an operator D in the following nonempty set:

$$\mathcal{S} = \{D \in Lip(X, X): \quad D^{-1} \text{ exists and is causal on } X$$
$$\text{such that } PD \in Lip(X, X)\}.$$

(2) Let $N = PD$.
(3) Choose an operator in the following family:

$$\mathcal{A} = \{A \in Lip(X, X): (I - AN)D^{-1} \in Lip(X, X)\}$$
$$\cap \{A \in Lip(X, X): \|AN\| < 1\}.$$

(4) Set $B = (I - AN)D^{-1}$.

Then, we obtain a right coprime factorization $P = ND^{-1}$ with a pair of operators (A, B) that satisfy the Bezout identity

$$AN + BD = I$$

on the Banach space X and hence stabilize the overall feedback system as desired.

§ 4. An Illustrative Example

For the purpose of illustrating the construction procedure described in the previous section, many examples can be easily given. In this section, we only give a very simple one (with less physical meaning), so that all detailed calculations can be precisely and completely carried out and explained.

Consider, therefore, an (SISO) feedback system shown as in Figure 5.2. We assume, in this feedback system, that $X = L_\infty = L_\infty([0, \infty))$ is the standard Banach space of real-valued measurable functions defined on $[0, \infty)$, with the associated extended linear space $X^e = L_\infty^e$.

Suppose that the plant operator P is given by the following simple (but nontrivial), unstable, and time-varying nonlinear operator:

$$P(\cdot)(t) = (t+1)I(\cdot)(t) + 1 : \quad L_\infty \to L_\infty^e \,,$$

where $I(\cdot)$ is the identity operator.

The set S defined in Theorem 5.6 is

$$S = \{D \in Lip(L_\infty, L_\infty) : \quad D^{-1} \text{ exists, causal on } L_\infty$$
$$\text{such that } PD \in Lip(L_\infty, L_\infty)\} \,.$$

It is clear that we have infinitely many choices in picking a D from S. Let us pick, for example,

$$D(\cdot)(t) = \frac{1}{t+1}I(\cdot)(t) \,.$$

Then, D is in $Lip(L_\infty, L_\infty)$ and

$$N(\cdot)(t) = PD(\cdot)(t) = I(\cdot)(t) + 1 \,,$$

which is also in $Lip(L_\infty, L_\infty)$. Moreover,

$$D^{-1}(\cdot)(t) = (t+1)I(\cdot)(t) \,,$$

which is unstable from L_∞ to L_∞. This is consistent with the fact that the given P is unstable. Thus, we have obtained a right factorization

$$P = (t+1)I(\cdot)(t) + 1 = ND^{-1} \,.$$

As pointed out in the previous section, we have infinitely many choices in finding an operator A such that the composite operator $(I - AN)D^{-1}$ is in $Lip(L_\infty, L_\infty)$. Again for simplicity we pick

$$A(\cdot)(t) = \left(1 - \frac{1}{t+1}\right) I(\cdot)(t).$$

Then, we have $A \in Lip(L_\infty, L_\infty)$, so that

$$AN(\cdot)(t) = \left(1 - \frac{1}{t+1}\right)\left[I(\cdot)(t) + 1\right],$$

and

$$(I - AN)(\cdot)(t) = \frac{1}{t+1}I(\cdot)(t) + \frac{1}{t+1} - 1.$$

To this end, we obtain

$$B(\cdot)(t) = (I - AN)D^{-1}(\cdot)(t) = I(\cdot)(t) + \frac{1}{t+1} - 1,$$

in which an unstable mode has been canceled. Moreover,

$$B^{-1} = I(\cdot) - \frac{1}{t+1} + 1.$$

Obviously, both B and B^{-1} are in $Lip(L_\infty, L_\infty)$.

Finally, it can be verified that A and B satisfy the Bezout identity. Indeed, we have

$$(AN + BD)(\cdot)(t)$$
$$= \left(1 - \frac{1}{t+1}\right)\left[I(\cdot)(t) + 1\right] + \left[\frac{1}{t+1}I(\cdot)(t) + \frac{1}{t+1} - 1\right]$$
$$= I(\cdot)(t)$$

as desired, so that the overall feedback system is stabilized.

§ 5. Left Coprime Factorization
for a Class of Nonlinear Control Systems

In this section, we turn to study the left coprime factorization problem for a class of general nonlinear control systems that are described by ordinary differential equations in the state-vector form. For such nonlinear control systems, we will not discuss the standard "feedback" configuration and will not take the Bezout identity into account. In this consideration, of course, the "coprimeness" must be redefined in an appropriate way. It will be seen, however, that this new formulation is equivalent to the classical one, as well as the one studied in the previous few sections, when the control system under investigation is linear.

We consider the following nonlinear control system:

$$\begin{cases} \dot{x}(t) = f(x(t), t) + g(u(t), t), \quad x(0) = x_0 \\ y(t) = h(x(t), t), \end{cases} \tag{5.15}$$

where $u(t)$ is an $m \times 1$ control input restricted in a (not necessarily bounded) subset $\Omega_u \subseteq L_{\tilde{p}}([0, \infty), R^m)$ of admissible controls, with $1 \leq \tilde{p} \leq \infty$ and R^n the image space of the vector-valued $L_{\tilde{p}}$-functions, $x(t)$ is the corresponding differentiable $n \times 1$ state vector in a (not necessarily bounded) subset $\Omega_x \subset L_p([0, \infty), R^n), 1 \leq p \leq \infty$ and usually $m \leq n$, with the initial state $x_0 \in \Omega_x$, $h(\cdot, t)$ is a vector-valued bounded operator defined on $\Omega_x \times [0, \infty)$, and the two nonlinear vector-valued mappings $f : \Omega_x \times [0, \infty) \to R^n$ and $g : \Omega_u \times [0, \infty) \to R^n$ are assumed to satisfy the following two conditions:

(a) $f(x, \cdot) : t \to f(x, t)$ and $g(u, \cdot) : t \to g(u, t)$ are integrable on $[0, \infty)$ for each $x \in \Omega_x$ and for each $u \in \Omega_u$, respectively.

(b) $f(\cdot, t) : x \to f(x, t)$ and $g(\cdot, t) : u \to g(u, t)$ are continuous on Ω_x and Ω_u, respectively, for almost all $t \in [0, \infty)$.

Observe that whenever we have an input-output relationship $x(t) = \phi(u)(t)$, we will always have $y(t) = H(\phi(u))(t)$ for a mapping H determined by the operator $h(\cdot, t)$. Hence, the operator $h(\cdot, t)$ is not significant in the following discussion, and so will be simply ignored below. In other words, we only consider the input-state behavior of the system in the following.

We also remark that the Banach spaces L_p and $L_{\tilde{p}}$ defined on the time domain $[0,\infty)$ here can certainly be replaced by the extended linear spaces consisting of functions belonging to $L_p([0,T], R^n)$ and $L_{\tilde{p}}([0,T], R^m)$ for all $T < \infty$, which was introduced in Chapter 4. It is easily seen that all results obtained below are valid for extended linear spaces. For simplicity of notation, we will not consider the extended setting in this section.

Now, by a direct integration of (5.15), we have

$$x(t) = x_0 + \int_0^t f(x(\tau), \tau)d\tau + \int_0^t g(u(\tau), \tau)d\tau \, .$$

Let

$$F(x)(t) = x_0 + \int_0^t f(x(\tau), \tau)d\tau \, , \tag{5.16}$$

$$N(u)(t) = \int_0^t g(u(\tau), \tau)d\tau \, , \tag{5.17}$$

and

$$D(x)(t) := (I - F)(x)(t) \, , \tag{5.18}$$

where I is the identity operator. Then, we obtain

$$D(x)(t) = N(u)(t) \, . \tag{5.19}$$

Assume, for the time being, that the nonlinear operator $(I - F)$ is invertible for almost all $t \in [0,\infty)$. Then, we have in a very natural way a *formal* left fractional representation for the nonlinear mapping from Ω_u to Ω_x, as follows:

$$x(t) = G(u)(t) := D^{-1}N(u)(t) \, . \tag{5.20}$$

We will study in the following the conditions under which the nonlinear operator D is invertible. To this end, we first notice that such formal left fractional representations are not unique in general since for any invertible operator T on $\mathcal{R}(N) \cap \mathcal{R}(D)$, where $\mathcal{R}(\cdot)$ is the range of the operator, we have $(TD)^{-1}(TN) = D^{-1}N$.

We next establish a useful result.

5.8. Theorem. *Suppose that $1 \leq p \leq \infty$ and that there exists a non-negative measurable function $c(\cdot) : [0, \infty) \to [0, \infty)$ satisfying $M < \infty$, where*

$$M = \begin{cases} \displaystyle\int_0^\infty c(\tau)d\tau & \text{for } p = \infty \\ \displaystyle\int_0^\infty \left(\int_0^t c(\tau)d\tau\right)^p dt & \text{for } 1 \leq p < \infty, \end{cases} \tag{5.21}$$

such that

$$|f(x_1, t) - f(x_2, t)| \leq c(t)\|x_1 - x_2\|_{L_p} \tag{5.22}$$

for all $(x_1, t), (x_2, t) \in \Omega_x \times [0, \infty)$, where $|\cdot|$ is the Euclidean norm of the vector-valued function. Then, the nonlinear operator F defined by (5.16), namely,

$$F(\cdot)(t) := x_0 + \int_0^t f(\cdot, \tau)d\tau,$$

is Lipschitz from L_p to itself. Consequently, the nonlinear operator $(I - F)(\cdot)(t)$ is Lipschitz in $Lip(\Omega_x)$. If, furthermore,

$$M < \begin{cases} 1 - \|F(0)\|_{L_p} & \text{for } p = \infty \\ (1 - \|F(0)\|_{L_p})^p & \text{for } 1 \leq p < \infty, \end{cases} \tag{5.23}$$

then $(I - F)(\cdot)(t)$ is invertible on Ω_x and $(I - F)^{-1} \in Lip(\Omega_x)$ with

$$\|(I - F)^{-1}\|_{Lip} \leq \|F(0)\|_{L_p} + (1 - \|F\|_{Lip})^{-1}. \tag{5.24}$$

Proof. Observe that for $p = \infty$, since

$$|F(x_1, t) - F(x_2, t)| \leq \int_0^t |f(x_1, \tau) - f(x_2, \tau)|d\tau$$

$$\leq \int_0^t c(\tau)\|x_1 - x_2\|_{L_\infty} d\tau$$

$$= \|x_1 - x_2\|_{L_\infty} \int_0^t c(\tau)d\tau,$$

by taking the essential supremum over $t \in [0, \infty)$ we have

$$\|F(x_1) - F(x_2)\|_{L_\infty} \leq M \|x_1 - x_2\|_{L_p} .$$

Hence, $F(\cdot)(t)$ is Lipschitz from L_∞ to itself. On the other hand, for $1 \leq p < \infty$, we have

$$|F(x_1, t) - F(x_2, t)|^p \leq \left(\int_0^t |f(x_1, \tau) - f(x_2, \tau)| d\tau \right)^p$$

$$\leq \left(\int_0^t c(\tau) \|x_1 - x_2\|_{L_p} d\tau \right)^p$$

$$= \|x_1 - x_2\|_{L_p}^p \left(\int_0^t c(\tau) d\tau \right)^p ,$$

so that

$$\|F(x_1) - F(x_2)\|_{L_p} \leq M^{1/p} \|x_1 - x_2\|_{L_p} ,$$

which implies that the nonlinear operator $F(\cdot)(t)$ is Lipschitz from L_p to itself.

If condition (5.23) is satisfied, then we have

$$\|F\|_{Lip} = \|F(0)\|_{L_p} + \sup_{\substack{x_1, x_2 \in \Omega_x \\ x_1 \neq x_2}} \frac{\|F(x_1) - F(x_2)\|_{L_p}}{\|x_1 - x_2\|_{L_p}} < 1 ,$$

so that Theorem 2.10 can be applied to conclude that $(I - F)(\cdot)(t)$ is invertible on Ω_x, no matter whether Ω_x is bounded or not in R^n, and $(I - F)^{-1} \in Lip(\Omega_x)$ with

$$\|(I - F)^{-1}\|_{Lip} \leq \|F(0)\|_{L_p} + (1 - \|F\|_{Lip})^{-1} .$$

This completes the proof of the theorem. $\square$

Similarly, we can establish the following result:

5.9. Theorem. *Suppose that* $1 \leq \tilde{p}, p \leq \infty$ *and that there exists a non-negative measurable function* $\tilde{c}(\cdot) : [0, \infty) \to [0, \infty)$ *satisfying* $\tilde{M} < \infty$, *where*

$$
\tilde{M} = \begin{cases} \displaystyle\int_0^\infty \tilde{c}(\tau) d\tau & \text{for } p = \infty \\[2ex] \displaystyle\int_0^\infty \left(\int_0^t \tilde{c}(\tau) d\tau \right)^p dt & \text{for } 1 \leq p < \infty, \end{cases} \tag{5.25}
$$

such that

$$
|g(u_1, t) - g(u_2, t)| \leq \tilde{c}(t) \|u_1 - u_2\|_{L_{\tilde{p}}} \tag{5.26}
$$

for all $(u_1, t), (u_2, t) \in \Omega_u \times [0, \infty)$. *Then, the nonlinear operator* N *defined by* (5.17), *namely,*

$$
N(\cdot)(t) := \int_0^t g(\cdot, \tau) d\tau,
$$

is Lipschitz from $L_{\tilde{p}}$ *to* L_p.

Note that under the assumptions of Theorems 5.8 and 5.9, all nonlinear operators N, D, D^{-1}, and $G = D^{-1}N$ are Lipschitz on their domains.

Next, we introduce the notion of the coprime factorization for the general nonlinear control system (5.15).

To start with, we need the concept of input-output stability for system (5.15). Let the input-output subsets Ω_u and Ω_x be prescribed and given, which are not necessarily bounded in the input-output spaces $L_{\tilde{p}}([0, \infty), R^m)$ and $L_p([0, \infty), R^n)$, respectively.

5.10. Definition. System (5.15) is said to be *input-output stable* (or simple *stable*) if the system operator G, defined implicitly by the operators F and N via (5.20), maps all the inputs from Ω_u into the subset Ω_x.

Hence, if the output subset Ω_x is strictly proper in the output space $L_p([0, \infty), R^n)$ and if the system operator G maps only some (but not all) inputs from Ω_u into Ω_x, or equivalently, if G maps some inputs from Ω_u into the nonempty subset $L_p([0, \infty), R^n)\backslash\Omega_x$, then the system is *unstable*.

In terms of the nonlinear operators D and N, system (5.15) is *stable* if $D(\Omega_x) \subseteq \Omega_x$, D^{-1} exists with $D^{-1}(\Omega_x) \subseteq \Omega_x$, and $N(\Omega_u) \subseteq \Omega_x$. In this case, we say that operators D, N, and D^{-1} are stable.

Note, as mentioned above, that by "input-output" we actually mean the "input-state" behavior of the system in this section.

5.11. Definition. System (5.15) is said to be *finite-gain stable* if the system operator G defined above is stable and moreover $\|G\|_{Lip} < \infty$.

Note that system (5.15) is finite-gain stable if the operators D, N, D^{-1} are Lipschitz and stable: $D(\Omega_x) \subseteq \Omega_x, N(\Omega_u) \subseteq \Omega_x, D^{-1}(\Omega_x) \subseteq \Omega_x$.

We will only consider finite-gain stability below. It is important to note that under the definition of stability given above, a bounded operator G with $\|G\|_{Lip} < \infty$ may not be stable since it may map an input from Ω_u into the set $L_p([0, \infty), R^n) \backslash \Omega_x$ (if it is nonempty).

Next, we are concerned with the causality. Recall that all generalized Lipschitz operators defined on extended linear spaces are causal. In this section, however, since we only consider the standard Banach space setting for the simplicity of notation, we have to assume the causality for the operators involved.

Now we define the left factorization for a nonlinear mapping from Ω_u to Ω_x by taking the system stability into account.

5.12. Definition. Let system (5.15) be given, which need not be stable. The corresponding nonlinear operator $G = D^{-1}N$ defined by (5.20) is a *left factorization* of the system over the space of stable operators if $N : \Omega_u \to \Omega_x$ and $D : \Omega_x \to \Omega_x$ are stable, D^{-1} exists (need not be stable) on Ω_x, and $DG = N$ on Ω_u.

Here, it is important to note that since we always require that the operator N be stable in the consideration of factorizations, the given nonlinear operator G is unstable if and only if the operator D^{-1} is unstable. This observation enables us to construct a left factorization for an unstable nonlinear control system, in which what we need is to guarantee the stability of the operators D and

N (which depends on the predesired input-output subsets Ω_u and Ω_x) and the existence of the inverse D^{-1} (a sufficient condition for this has been established in Theorem 5.8 above). Hence, if the given system (5.15) satisfies all the conditions stated in Theorems 5.8 and 5.9, and if the input-output subsets Ω_u and Ω_x are given such that $N(\Omega_u) \subseteq \Omega_x$ and $D(\Omega_x) \subseteq \Omega_x$, then we have actually constructed a left factorization for it using the formula (5.20).

We next consider the coprimeness.

5.13. Definition. Let system (5.15) be given, which need not be stable. The corresponding nonlinear operator $G = D^{-1}N$ defined by (5.20) is a *left coprime factorization* of the system over the space of stable operators if $D^{-1}N$ is a left factorization and moreover there exist two constants $\alpha \geq 0$ and $\beta \geq 0$, not both zero if $m \geq n$ and both nonzero if $m < n$, where m and n are dimensions of the input and state vectors u and x, respectively, such that

$$
\begin{cases}
\|Nu\|_{L_p} \geq \alpha \|u\|_{L_{\tilde{p}}} & \text{for all } u \in \Omega_u \\
\|Dx\|_{L_p} \geq \beta \|x\|_{L_p} & \text{for all } x \in \Omega_x .
\end{cases}
\tag{5.27}
$$

A few remarks are in order. First, since the set Ω_u of admissible control inputs may not be compact, in fact it might be a normed linear space, and since the mapping $N : \Omega_u \to \Omega_x$ may not be injective here, the theory of left coprime factorization studied below is not a simple issue. The qualitative meaning of the coprimeness defined above is that for any unbounded input u yielding an unbounded output $x = Gu$ through the operator G, either Nu or Dx is unbounded.

Secondly, as remarked above for the formal left factorization, a left coprime factorization is not unique either, since for any stable operator T on $\mathcal{R}(N) \cap \mathcal{R}(D)$ with a stable inverse T^{-1}, we have $(TD)^{-1}(TN) = D^{-1}N$. Conversely, if $G = D_1^{-1}N_1 = D_2^{-1}N_2$, then there is a stable T on $\mathcal{R}(N_1) \cap \mathcal{R}(D_1)$ with a stable inverse T^{-1} such that $N_2 = TN_1$ and $D_2 = TD_1$. Indeed, this T is given by $T = D_2 D_1^{-1}$.

Finally, we point out that for linear systems, this coprimeness is equivalent to that from the algebraic approach. To illustrate this, we first recall that in the algebraic approach in linear systems theory, a given $m \times n$ proper rational matrix $G(s)$ has left co-prime factorization over the ring of proper stable rational matrices if $G(s) = D^{-1}(s)N(s)$, where

(i) $D(s)$ and $N(s)$ are $m \times m$ and $m \times n$ proper stable rational matrices, respectively, with det $D(s) \not\equiv 0$, and

(ii) there exist an $n \times m$ matrix $X(s)$ and an $m \times m$ matrix $Y(s)$, both proper rational and stable, such that

$$N(s)X(s) + D(s)Y(s) = I. \tag{5.28}$$

Then, we show the equivalence as follows. First, consider the direction if the above conditions hold. Since $N(s)$ and $D(s)$ are proper rational and stable, the corresponding linear mappings $N(\cdot)$ and $D(\cdot)$ are causal and finite-gain stable. Moreover, since $det D(s) \not\equiv 0$, D^{-1} exists, so that the inverse linear mapping $D^{-1}(\cdot)$ exists and is bounded. Hence, we have a left fractional representation $D^{-1}N$ over the space of finite-gain stable operators. To prove the coprimeness, namely, to verify condition (5.27), we observe that condition (5.28) implies that

$$rank[N(s) \quad D(s)] = m$$

for all s on the right-half complex plane including the extended imaginary axis, namely, the matrix $[N(s) \quad D(s)]$ has no zero in the right-half extended complex plane, $Re(s) \geq 0$ including the point at infinity. Consequently, we have

$$\inf_{-\infty \leq \omega \leq \infty} \lambda_{min}\left([N(j\omega) \quad D(j\omega)]\right) > 0$$

where $\lambda_{min}(\cdot)$ is the smallest eigenvalue of the argument matrix, so that condition (5.27) is satisfied. The converse direction can be verified by retracing the above procedure.

It must be pointed out that when the extended linear space setting is considered, the verification of the coprimeness requires a little more effort since we have to show, in addition, that there exist no functions u and x belonging to the nontrivial extended linear spaces (which are not regular Banach spaces) such that $N(u)$ and $D(x)$ are still in the regular Banach spaces. We are not getting into the detailed investigation for this issue here.

In what follows, we will establish a sufficient condition for system (5.15) to have a left coprime factorization by imposing restrictions on the two nonlinear mappings $f(\cdot, t)$ and $g(\cdot, t)$. We have given in Theorem 5.8 sufficient conditions on f and g for the two nonlinear operators D and N to be Lipschitz. Observe, moreover, that Lipschitz operators are both bounded and continuous. Hence, what we need is basically to satisfy condition (5.27). For this purpose, we show the following result.

5.14. Theorem. *Suppose that all the conditions stated in Theorems 5.8 and 5.9 are satisfied. In addition, suppose that the following two conditions hold true:*

(i) *There exists a constant $W : 0 \leq W \leq 1$, such that for $p = \infty$,*

$$\operatorname*{ess\ sup}_{0 \leq t < \infty} \left| x_0 + \int_0^t f(x, \tau) d\tau \right| \leq W \|x\|_{L_\infty}$$

for all $(x, t) \in \Omega_x \times [0, \infty)$, and for $p = 1$,

$$\int_0^\infty \left| x_0 + \int_0^t f(x, \tau) d\tau \right| dt \leq W \|x\|_{L_1}$$

for all $(x, t) \in \Omega_x \times [0, \infty)$; and there exists a constant $W : 0 \leq W < \infty$, such that for $1 < p < \infty$,

$$\int_0^\infty \left| x(t) - x_0 - \int_0^t f(x, \tau) d\tau \right|^p dt \geq W \|x\|_{L_p}^p$$

for all $(x, t) \in \Omega_x \times [0, \infty)$.

(ii) *There exists a constant $\tilde{W} : 0 \leq \tilde{W} < \infty$, such that for $p = \infty$,*

$$\operatorname*{ess\ sup}_{0 \leq t < \infty} \left| \int_0^t g(u, \tau) d\tau \right| \geq \tilde{W} \|u\|_{L_{\tilde{p}}}$$

for all $(u, t) \in \Omega_u \times [0, \infty)$, and for $1 \leq p < \infty$,

$$\int_0^\infty \left| \int_0^t g(u, \tau) d\tau \right|^p dt \geq \tilde{W} \|u\|_{L_{\tilde{p}}}^p$$

for all $(u, t) \in \Omega_u \times [0, \infty)$. Here, it is assumed that not both conditions (a) and (b) listed below hold simultaneously if $m \geq n$ and that both (a) and (b) do not hold if $m < n$, where m and n are the dimensions of the input and state vectors u and x, respectively:

(a) $W = 1$ (for $p = 1$ or ∞) or $W = 0$ (for $1 < p < \infty$);

(b) $\tilde{W} = 0$.

Moreover, assume that for the prescribed input-output subsets Ω_u and Ω_x, the operators D and N defined in (5.17) and (5.18) satisfy $D(\Omega_x) \subseteq \Omega_x$ and $N(\Omega_u) \subseteq \Omega_x$, respectively. Then, the formal left factorization (5.20) is a left coprime factorization over the space of stable operators for the nonlinear input-output mapping $G : \Omega_u \rightarrow \Omega_x$ that describes the nonlinear control system (5.15).

Proof. For $p = \infty$, we have

$$|D(x)(t)| = \left| I(x)(t) - x_0 - \int_0^t f(x, \tau) d\tau \right|$$

$$\geq |x(t)| - \left| x_0 + \int_0^t f(x, \tau) d\tau \right| ,$$

so that by taking the essential supremum over $t \in [0, \infty)$,

$$\|D(x)\|_{L_\infty} \geq \|x\|_{L_\infty} - W\|x\|_{L_\infty} := \beta\|x\|_{L_\infty} ,$$

where $0 \leq \beta := 1 - W$, and

$$\|N(u)\|_{L_\infty} = ess \sup_{0 \leq t < \infty} \left| \int_0^t g(u, \tau) d\tau \right|$$

$$\geq \tilde{W}\|u\|_{L_{\tilde{p}}} := \alpha\|u\|_{L_{\tilde{p}}} ,$$

where $0 \leq \alpha := \tilde{W} < \infty$. The conditions (a) and (b) on W and $\tilde{W}$ guarantee that the requirements on the constants α and β and hence the conditions in (5.27) are all satisfied.

For $p = 1$, the above procedure can be carried out in the same manner.

For $1 < p < \infty$, we have

$$\|D(x)\|_{L_p} = \left(\int_0^\infty \left| x(t) - x_0 - \int_0^t f(x,\tau)d\tau \right|^p dt \right)^{1/p}$$

$$\geq W^{1/p}\|x\|_{L_p} := \beta \|x\|_{L_p},$$

where $0 \leq \beta := W^{1/p} < \infty$. Moreover, we have

$$\|N(u)\|_{L_p} = \left(\int_0^\infty \left| \int_0^t g(u,\tau)d\tau \right|^p dt \right)^{1/p}$$

$$\geq \tilde{W}^{1/p}\|u\|_{L_{\tilde{p}}} := \alpha \|u\|_{L_{\tilde{p}}},$$

where $0 \leq \alpha := \tilde{W}^{1/p} < \infty$. Again, the conditions (a) and (b) on W and $\tilde{W}$ guarantee that the requirements on the constants α and β and hence the conditions in (5.27) are all satisfied. Finally, the assumptions $D(\Omega_x) \subseteq \Omega_x$ and $N(\Omega_u) \subseteq \Omega_x$ guarantee the stability of the operators D and N on the prescribed input-output subsets Ω_u and Ω_x.   □

Next, for the purpose of illustration, let us show a very simple example of a single-input/single-output (SISO) nonlinear control system. Although many examples can be easily given, it is certainly interesting to see one with an unstable system operator.

5.15. Example. Consider the nonlinear system

$$\begin{cases} \dot{x}(t) = a(t)\sin(x(t)) + b(t)u(t) \\ x(0) = 0, \end{cases}$$

where it is assumed that
(1) $\int_0^\infty |a(t)|dt < 1$;
(2) $\int_0^\infty |b(t)|dt \leq 1$.
For this system, we consider the input-output subsets and spaces:

$$\Omega_u = \Omega_x = L_{\tilde{p}} = L_1 \cap L_\infty \quad \text{and} \quad L_p = L_\infty,$$

where all the notation is as above. As indicated before, if the system
operator maps an input from Ω_u to the (nonempty) subset $L_\infty \backslash \Omega_x$,
then the system is unstable.

We first verify that all conditions stated in Theorems 5.8, 5.9,
and 5.14 are satisfied. Indeed, we have

$$
\begin{aligned}
&|f(x_1, t) - f(x_2, t)| \\
&= |a(t) \sin(x_1(t)) - a(t) \sin(x_2(t))| \\
&\leq |a(t)| \, \|x_1 - x_2\|_{L_\infty} ,
\end{aligned}
$$

so that we may simply choose $c(t) = |a(t)|$. Thus, since $F(0)(t) \equiv 0$,
we have

$$
M = \int_0^\infty c(\tau) d\tau = \int_0^\infty |a(t)| dt < 1 - \|F(0)\|_{L_\infty} ,
$$

and moreover,

$$
\begin{aligned}
&\operatorname*{ess\,sup}_{0 \leq t < \infty} \left| \int_0^t a(\tau) \sin(x(\tau)) d\tau \right| \\
&= \operatorname*{ess\,sup}_{0 \leq t < \infty} \left| \int_0^t a(\tau) \frac{\sin(x(\tau))}{\|x\|_{L_\infty}} d\tau \right| \|x\|_{L_\infty} \\
&\leq W \|x\|_{L_\infty} ,
\end{aligned}
$$

where

$$
\begin{aligned}
W &:= \sup_{x \in \Omega_x} \operatorname*{ess\,sup}_{0 \leq t < \infty} \left| \int_0^t a(\tau) \frac{\sin(x(\tau))}{\|x\|_{L_\infty}} d\tau \right| \\
&\leq \int_0^\infty |a(\tau)| d\tau < 1 .
\end{aligned}
$$

The conditions on $b(t)$ can be similarly verified, which are in fact
much simpler and imply that $N(\Omega_u) \subseteq \Omega_x$. Hence, all conditions
stated in Theorems 5.8, 5.9, and 5.14 are satisfied.

Thus, we have actually constructed a left coprime fractional
representation using Formula (5.20) for the given system:

$$
G(u)(t) = D^{-1} N(u)(t) ,
$$

with

$$D(x)(t) = (I - F)(x)(t) = x(t) - \int_0^t a(\tau)\sin(x(\tau))d\tau\,,$$

and

$$N(u)(t) = \int_0^t b(\tau)u(\tau)d\tau\,,$$

where N, D, D^{-1} are all Lipschitz by Theorems 5.8 and 5.9.

A numerical example of an unstable system of this type is given by $a(t) = (t-1)e^{-t}$ with $b(t) = e^{-t}$. To see this, we first observe that

$$\int_0^\infty |a(\tau)|d\tau = \int_0^1 (1-\tau)e^{-\tau}d\tau + \int_1^\infty (\tau-1)e^{-\tau}d\tau = \frac{2}{e} < 1\,,$$

and that $\int_0^\infty |b(\tau)|d\tau = 1$, so that conditions (1) and (2) assumed above are satisfied. Moreover, we have

$$\left\| \int_0^t b(\tau)u(\tau)d\tau \right\|_{L_1} \leq \int_0^\infty |e^{-\tau}|d\tau \cdot \|u\|_{L_1} = \|u\|_{L_1}\,,$$

which implies that $N(\Omega_u) \subseteq \Omega_x$, namely, the operator N is stable. The stability of the operator D can be similarly verified.

Now, the input signal $u(t) = e^{-t/3}$ is in the input subset $\Omega_u = L_1 \cap L_\infty([0, \infty), R^1)$, but the output $x(t)$ satisfies

$$x(t) = \int_0^t (\tau-1)e^{-\tau}\sin(x(\tau))d\tau + \int_0^t e^{-\tau}e^{-\tau/3}d\tau$$

$$\geq \int_0^t e^{-4\tau/3}d\tau - \int_0^\infty |(\tau-1)e^{-\tau}|d\tau$$

$$\geq \frac{3}{4} - \epsilon - \frac{2}{e}$$

$$> 0\,,$$

for any fixed $\epsilon \in (0, \frac{3}{4} - \frac{2}{e})$ and for all large enough values of t, which implies that $x(t)$ is not in the output subset $\Omega_x = L_1([0, \infty), R^1)$. However, it can also be seen from the above that

$$\|x\|_{L_\infty} \leq \frac{3}{4} + \frac{2}{e}\,,$$

which implies that $x(t) \in L_\infty([0, \infty), R^1)$. Hence, the system under investigation is unstable.

§ 6. More on the Left Coprime Factorization

We have discussed, in the above, how to actually construct a left coprime factorization for a (stable or unstable) nonlinear control system described by the vector-valued nonlinear differential equation (5.15). We close this chapter by the following important remark, which relates this state-vector approach to the feedback configurations discussed in detail in the previous chapters. Consider, therefore, the general feedback control system shown in Figure 5.3, in which P is the plant operator, C the compensator, and N a filter, all need not be linear, and u is the control input with y the system output.

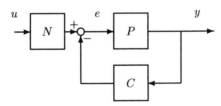

Figure 5.3.   A nonlinear feedback system.

If we are concerned with the error signal e, then we have the following relationship:

$$(I + CP)(e) = N(u),\qquad\qquad (5.29)$$

which can be easily verified from the configuration. Hence, under the general framework developed in this chapter, by assuming that the nonlinear operators C and N are Lipschitz and by designing the compensator C to be such that the composite operator CP is also Lipschitz with $\|CP\|_{Lip} < 1$, we see that the nonlinear operator $(I + CP)$ is invertible with a Lipschitz inverse $(I + CP)^{-1}$. Consequently, it follows from (5.29) that

$$e = (I + CP)^{-1}N(u),$$

which has yielded a formal left factorization for the nonlinear mapping from the input to the error-output. To this end, the basic idea and technique shown in the last section can be employed to develop the notion of coprimeness for this nonlinear mapping for the nonlinear feedback system described in Figure 5.3.

Exercises

5.1. Verify that the set $Lip(X, Y)$ defined in (5.4) is indeed an infinite-dimensional Banach space.

5.2. Formulate the left coprime factorization counterpart of the right coprime factorization, Problem 5.5.

5.3. Show that a one-to-one (i.e., injective) and onto (i.e., surjective) operator has a two-sided inverse.

5.4. Imitating 5.7, derive sufficient conditions for constructing a left coprime factorization for the nonlinear feedback system shown in Figure 5.2. Following the construction of the example shown in Section 4, give an example to show the construction of a left coprime factorization of an unstable (yet stabilizable) and time-varying nonlinear feedback control system.

5.5. Supply a proof for Theorem 5.9.

5.6. Complete the discussions given in Section 6 by supplying detailed derivations for the construction of a left coprime factorization of the nonlinear control system shown in Figure 5.3. Give as well a concrete example to show the construction.

References

[1] G. Chen, "A note on the coprime fractional representation of nonlinear feedback systems," *Sys. Contr. Lett.*, **14** (1990), 41– 43.

[2] G. Chen and R..J.P. de Figueiredo, "Construction of the left coprime fractional representation for a class of nonlinear control systems," *Sys. Contr. Lett.*, **14** (1990), 353–361.

[3] G. Chen and R.J.P. de Figueiredo, "On construction of coprime factorizations of nonlinear feedback control systems," *Circ. Sys. Sign. Proc.*, **11** (1992), 285–307.

[5] R. Danow and G. Chen, "A necessary and sufficient condition for right coprime factorization of nonlinear feedback control systems," *Circ. Sys. Sign. Proc.*, **12** (1993), 489–492.

[6] C.A. Desoer and R.W. Liu, "Global parametrization of feedback systems," *Sys. Contr. Lett.*, **1** (1982), 249–251.

[7] C.A. Desoer and M.G. Kabuli, "Right factorizations of a class of time-varying nonlinear systems," *IEEE Trans. Auto. Contr.*, **33** (1988), 755–757.

[8] J. Hammer, "Nonlinear systems: stability and rationality," *Int. J. Contr.*, **40** (1984), 1–35.

[9] J. Hammer, "Fraction representations of nonlinear systems: a A simplified approach," *Int. J. Contr.*, **46** (1987), 455–472.

[10] E.D. Sontag, "Smooth stabilization implies coprime factorization," *IEEE Trans. Auto. Contr.*, **34** (1989), 435–443.

[11] E.D. Sontag, "Some connections between stabilization and factorization," *Proc. of Contr. Decis. Conf.*, Tampa, FL, 1989.

[12] T.T. Tay and J.B. Moore, "Left coprime factorizations and a class of stabilizing controllers for non-linear systems," *Int. J. Contr.*, **49** (1989), 1235–1248.

[13] M.S. Verma, "Coprime fractional representations and stability of non-linear feedback systems," *Int. J. Contr.*, **48** (1988), 897– 918.

[14] M.S. Verma and L.R. Hunt, "Right coprime factorizations and stabilization for nonlinear systems," *IEEE Trans. Auto. Contr.*, **38** (1993), 222–231.

Chapter 6

Nonlinear System Identification

The system identification problem arises in control systems theory when very little information on the plant to be controlled is available *a priori*, or when, as in the case of adaptive control systems, the structure of the plant is changing in an unknown manner as a function of time. From now on, by "a system to be identified" we mean a plant appearing in situations like the ones mentioned.

System identification may be defined as the problem of determining a precise mathematical description of the system under consideration based on some *a priori* knowledge of the system and/or on some data representative of its input-output behavior. The *a priori* knowledge available may be in terms of the particular physical makeup of the system and some of the laws or rules governing it or, more generally, in terms of the properties of a class to which the system may be known to belong. In what follows, we will characterize this class by specifying the spaces and sets to which the system inputs, outputs, and input-output maps belong. With regard to available input-output data, we will assume that they are in the form of samples of a set of representative inputs and corresponding outputs.

There is a huge volume of literature on system identification, mostly on linear system identification. In keeping with the concepts presented in this book, we will focus our attention on two issues. One is the theoretical issue of information-based complexity of the nonlinear system identification problem in the worst-case setting. This is discussed in the first three sections. The other is that of the recovery of the input-output description of a system, in terms of a general nonlinear Nth-order difference equation in a generalized Fock space setting. This is studied in Sections 4 and 5.

Specifically, we state the information-based complexity problem under consideration in Section 1 and then establish both lower and upper bounds for the identification-error functional in Section 2. We then provide in Section 3 an existence result for an optimal algorithm for the identification, which achieves the infimum of the error. In Section 4, as a prelude to the developments in Section 5, we introduce the generalized Fock space $F(E^N)$ on a Euclidean space E^N whose members are Volterra series in N variables. Next, in Section 5, we assume that the input-output samples of the system to be identified are related to each other by a general Nth-order nonlinear difference equation, the left-hand side of which is a member of $F(E^N)$. This in turn permits us to retrieve this member, and hence the difference equation, by either solving a min-max problem or using least squares inversion in $F(E^N)$, yielding a unique solution to the nonlinear system identification problem under the assumed conditions. Finally in Section 6, we discuss very briefly how these identification techniques can be extended to continuous-time-parameter systems.

§ 1. Problem Formulation

In this section, we formulate the so-called *information-based complexity* problem for optimal identification of general nonlinear control systems.

Consider the general nonlinear system shown in Figure 6.1.

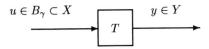

Figure 6.1.   A nonlinear input-output system.

We assume the following conditions:
(1) A Banach function space X and a closed ball B_γ of radius $\gamma > 0$ in X are given; and B_γ is assumed to constitute the admissible class of bounded inputs u of the system.
(2) We assume that the corresponding system outputs $y = T(u)$ belong to another given Banach function space Y, in which $T : B_\gamma \to Y$ is an unknown but well-defined and bounded nonlinear operator to be identified. Here, both u and y may be vector-valued.
(3) Let Z denote the product space of two normed linear spaces, such as $Z_1 \otimes Z_2$, Z_1 being n-dimensional and Z_2 m-dimensional normed linear function spaces respectively, each component of which belongs to L_p, say. Suppose that a data vector $z \in Z$ (not necessarily accurate) about the system input, u, and system output, y, is available. More precisely, suppose that z is obtained from a sampling operator $S : B_\gamma \otimes T(B_\gamma) \to Z$ with a possible error $\|z - S(u, y)\|_Z \le \epsilon$, where $y = T(u)$, $\epsilon \ge 0$, and S is assumed to be linear, bounded, and surjective. This may be described as

$$z \in S(B_\gamma, T(B_\gamma)) + \epsilon U$$

with U the unit ball in Z. In the case that the input data on u is not available, it is understood that S maps B_γ to an empty set. When $\epsilon = 0$, the data information is accurate.

(4) Let Θ be the family of all possible *algorithms* (which need not be linear nor realistic) $\theta : Z \to Y$, where θ maps the data $z \in Z$ to an element $\theta(z) \in Y$, which can be used as an approximation of the actual system output $y = T(u) \in Y$. The family Θ can be precisely specified for a given practical problem. In the most general case, Θ may be assumed to be the family of all possible algorithms that map the data z into the space Y.

The relationships among all these entities are shown in Figure 6.2.

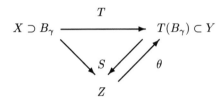

Figure 6.2.   Relations among several nonlinear mappings.

Once an algorithm θ has been determined under a certain criterion, the restriction of the operator $\theta(S(\cdot,\cdot))$, namely, $\theta(S(\cdot,\cdot))\big|_{B_\gamma}$, on the set B_γ (which is the first component of the domain) can be used as a reconstruction (identification) of the unknown system mapping $T(\cdot)$.

In the investigation of complexity in this section, however, we are not concerned with what method to use in finding such an algorithm and how to actually construct the algorithm. Instead, we study the following issue: Define an *identification-error functional*:

$$e(\theta) := \sup_{\substack{u \in B_\gamma \\ \|S(u,T(u))-z\|_Z \le \epsilon}} \|T(u) - \theta(z)\|_Y . \tag{6.1}$$

The objective is to study the following *information-based complexity* problem:

6.1. Problem.   *Does there exist an algorithm* $\theta^* = \theta^*(z)$ *in* Θ, *based on the given data-information*

$$z \in S(B_\gamma, T(B_\gamma)) + \epsilon U \subset Z ,$$

that minimizes the error $e(\theta, \epsilon)$? If so, what are possible lower and upper bounds for this error when the optimal algorithm θ^* is applied to solve the system identification problem?

Here, an optimal algorithm θ^* is one in Θ that yields the following minimum identification error:

$$e(\theta^*) = \inf_{\theta \in \Theta} e(\theta) . \tag{6.2}$$

We remark once again that the above mathematical setting is in general for multi-input/multi-output (MIMO) nonlinear control systems, since u and y may be vector-valued. In the next two sections, we prove the existence of an optimal algorithm $\theta^* \in \Theta$ (usually not unique) under mild conditions and establish both lower and upper bounds for the identification error $e(\theta)$, while in section 5 we present an actual solution in a generalized Fock space setting.

§ 2. Lower and Upper Bounds

In this section, we first establish a lower and an upper bound for the identification error $e(\theta)$ defined in (6.1). The existence result for an optimal algorithm will be given in the next section. Let

$$\underline{e} = \inf_{\theta \in \Theta} e(\theta).$$

Then we have the following lower bound for the error.

6.2. Theorem. *A lower bound for the identification error $e(\theta)$ is given by*

$$\underline{e} \geq \sup_{\substack{\|S(u_i, T(u_i)) - z\|_Z \leq \epsilon \\ u_i \in B_\gamma, i=1,2}} \frac{1}{2} \|T(u_1) - T(u_2)\|_Y. \qquad (6.3)$$

We remark that for the constant (nonlinear) operator $T(\cdot) \equiv c$, when $S = I$ (the identity operator) with $\epsilon = 0$, the algorithm $\theta(z) = z$ is optimal as it achieves the lower bound $\underline{e} = 0$.

Proof. The result follows easily from the following inequalities:

$$\frac{1}{2} \|T(u_1) - T(u_2)\|_Y$$
$$\leq \frac{1}{2} \|T(u_1) - \theta(z)\|_Y + \|T(u_2) - \theta(z)\|_Y$$
$$\leq \max_{i=1,2} \|T(u_i) - \theta(z)\|_Y.$$

Taking supremum over $u_1, u_2 \in B_\gamma$ subject to

$$\|S(u_i, T(u_i)) - z\|_Z \leq \epsilon, \qquad i = 1, 2,$$

and using (6.1) yields the lower bound and completes the proof of the theorem. $\qquad \square$

Next, we establish an upper bound for the identification error $e(\theta)$. Define

$$\bar{e} = \sup_{\theta \in \Theta} e(\theta).$$

Then, the following is an estimate for the upper bound.

6.3. Theorem. *An upper bound for the identification error $e(\theta)$ is given by*

$$\bar{e} \leq \gamma(2 + \epsilon)\|T\|_0, \qquad (6.4)$$

where

$$\|T\|_0 := \sup_{0 \neq u \in B_\gamma} \frac{\|T(u)\|_Y}{\|u\|_X}.$$

We remark that if the bounded nonlinear operator $T : B_\gamma \to Y$ has a uniformly convergent series representation, a convergent Volterra expansion studied in Section 9 of Chapter 1, for example, in the form of

$$T(\cdot) = \sum_{n=0}^{N} f_n(\cdot),$$

where $N \leq \infty$ and f_n is an n-linear operator, then the bound $\|T\|_0$ shown above can be improved to be the following:

$$\|T\| := \|T(0)\|_Y + \sup_{\substack{u,v \in B_\gamma \\ u \neq v}} \frac{\|(T - T(0))(u) - (T - T(0))(v)\|_Y}{\|u - v\|_X}, \qquad (6.5)$$

which is a finite number with $T(0) = f_0$.

Proof. First, we notice that since $T : B_\gamma \to Y$ is a bounded operator, $T(B_\gamma)$ is a bounded subset in Y. Because S is a surjective mapping from $B_\gamma \otimes T(B_\gamma)$ onto Z, for any $z \in U$ there exists at least one $\tilde{u}$ in B_γ such that $S(\tilde{u}, T(\tilde{u})) = z$. Set

$$G(z) = \{\, u \in B_\gamma \; : \; S(u, T(u)) = z \,\},$$

and consider the intersection of $G(z)$ and cB_γ, where

$$cB_\gamma := \{u \in X : \quad \|u\|_X = c\gamma, \; 0 < c \leq 1\},$$

as shown in Figure 6.3.

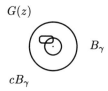

Figure 6.3. Relations among several sets.

Define
$$c^*(z) = \inf \{\, c : \quad G(z) \cap cB_\gamma \neq \phi \,\}.$$
Then, we have
$$c^*(z)\gamma = \inf\{\|u\|_X : \quad u \in G(z)\},$$
or
$$c^*(z) = \gamma^{-1} \cdot \inf\{\|u\|_X : u \in G(z)\}.$$
Now, for any $u \in B_\gamma$ and $u^0 \in G(z)$, since $S(u + \epsilon u^0, T(u + \epsilon u^0))$
is an element in Z and since S is surjective, it follows that for any
$\mu > 0$ there exists a u^* in the set
$$G(S(u + \epsilon u^0, T(u + \epsilon u^0))) \cap c_0 B_\gamma$$
for some c_0 satisfying
$$0 < c_0 < c^*(S(u + \epsilon u^0, T(u + \epsilon u^0))) + \mu,$$
such that
$$S(u^*, T(u^*)) = S(u + \epsilon u^0, T(u + \epsilon u^0)).$$
Moreover, we have
$$
\begin{aligned}
c_0 &< c^*(S(u + \epsilon u^0, T(u + \epsilon u^0))) + \mu \\
&= \gamma^{-1} \cdot \inf\{\, \|\tilde{u}\|_X : \quad \tilde{u} \in G(S(u + \epsilon u^0, T(u + \epsilon u^0))) \,\} + \mu \\
&\le \gamma^{-1}\|u + \epsilon u^0\|_X + \mu \\
&\le \gamma^{-1}(\gamma + \epsilon\gamma) + \mu \\
&\le 1 + \epsilon + \mu.
\end{aligned}
$$

For any algorithm

$$\theta : \ S(u + \epsilon u^0, T(u + \epsilon u^0)) \to T(u^*),$$

it follows from (6.1) that

$$
\begin{aligned}
e(\theta) &= \sup_{\substack{u \in B_\gamma \\ \|S(u,T(u))-z\|_Z \leq \epsilon}} \|T(u) - \theta(z)\|_Y \\
&= \sup_{\substack{u \in B_\gamma \\ \|z\|_Z \leq 1}} \|T(u) - \theta(S(u,T(u)) + \epsilon z)\|_Y \\
&= \sup_{\substack{u \in B_\gamma \\ u^0 \in G(z)}} \|T(u) - \theta(u + \epsilon u^0, S(u + \epsilon u^0, T(u + \epsilon u^0)))\|_Y \\
&= \sup_{\substack{u \in B_\gamma \\ u^* \in c_0 B_\gamma}} \|T(u) - T(u^*)\|_Y \\
&\leq \|T\|_0 \sup_{\substack{u \in B_\gamma \\ u^* \in c_0 B_\gamma}} (\|u\|_X + \|u^*\|_X) \qquad\qquad (*) \\
&\leq (\gamma + \gamma(1 + \epsilon + \mu))\|T\|_0 \, .
\end{aligned}
$$

Since μ is arbitrary, we have $\bar{e} \leq \gamma(2 + \epsilon)\|T\|_0$. $\qquad\qquad\square$

To verify (6.5), we simply observe that under the assumption on T, inequality $(*)$ obtained above can be improved as follows:

$$
\begin{aligned}
e(\theta) & \\
&\leq \sup_{\substack{u \in B_\gamma \\ u^* \in c_0 B_\gamma}} \|T(0)(u - u^*) + (T - T(0))(u) - (T - T(0))(u^*)\|_Y \\
&\leq \|T\| \sup_{\substack{u \in B_\gamma \\ u^* \in c_0 B_\gamma}} (\|u\|_X + \|u^*\|_X) \\
&\leq \gamma(2 + \epsilon + \mu)\|T\| \, .
\end{aligned}
$$

Since μ is arbitrary, it follows that $\bar{e} \leq \gamma(2 + \epsilon)\|T\|$.

§ 3. Existence of Optimal Algorithms

Now, we establish the existence of an optimal algorithm $\theta^* \in \Theta$ for the information-based complexity problem discussed in the first section.

We first need some new notation. In addition to all the notations used in the last section, for a fixed element $z \in Z$ we denote by H_z a subset in the Banach space Y defined by

$$H_z = \{T(u) : \ u \in B_\gamma, \ \|S(u, T(u)) - z\|_Z \le \epsilon\}. \tag{6.7}$$

It is clear that $H_z \ne \phi$ if and only if

$$z \in \{ \ S(B_\gamma, T(B_\gamma)) + \epsilon U \ \},$$

where again U is the unit ball in Z. It then follows from (3) in the system set-up (in Section 1) that $H_z \ne \phi$.

6.4. Definition. An element $y_0 \in Y$ is called a *Chebyshev center* for the set H_z if

$$\sup_{v \in H_z} \|y_0 - v\|_Y = \inf_{y \in Y} \sup_{v \in H_z} \|y - v\|_Y, \tag{6.8}$$

where

$$r(H_z) := \inf_{y \in Y} \sup_{v \in H_z} \|y - v\|_Y$$

is called the *Chebyshev radius* of the set H_z.

Note that a Chebyshev center of a set is not necessarily an element of the set and may not be unique.

Now, we are in a position to state and prove the following existence result:

6.5. Theorem. *Suppose that the Banach space Y is reflexive. Then a Chebyshev center for the subset H_z defined in (6.7), denoted by $y_0 = y_0(z)$, exists. Consequently, any algorithm $\theta^* \in \Theta$ with*

$$\theta^* : \quad z \to y_0$$

is optimal, and we have

$$\bar{e} = \sup_{z \in \{S(B_\gamma, T(B_\gamma)) + \epsilon U\}} r(H_z) . \qquad (6.9)$$

Proof. First, since $z \in \{S(B_\gamma, T(B_\gamma)) + \epsilon U\}$, the set $H_z \neq \phi$ as remarked above. Then, since $T : B_\gamma \to Y$ is bounded, the set H_z is a bounded subset in $T(B_\gamma)$. Note, moreover, that Y is a reflexive Banach space to which the set H_z belongs. Hence, by a fact well known in approximation theory (see Reference [5]), a Chebyshev center for H_z exists that is the center of a ball with minimum radius that contains the closure of H_z in Y.

To prove the second part of the theorem, let θ be an arbitrary algorithm $\theta \in \Theta$. Then, for all $u \in B_\gamma$ satisfying

$$\|S(u, T(u)) - z\|_Z \leq \epsilon ,$$

we have

$$\|T(u) - \theta(z)\|_Y \leq e(\theta) .$$

It follows that

$$\sup_{v \in H_z} \|y - v\|_Y \leq e(\theta) ,$$

where $y = \theta(z)$ and $v \in H_z$. Consequently, we have

$$r(H_z) \leq e(\theta) ,$$

so that

$$\sup_{z \in \{S(B_\gamma, T(B_\gamma)) + \epsilon U\}} r(H_z) \leq \sup_{\theta \in \Theta} e(\theta) = \bar{e} .$$

On the other hand, we observe that

$$\|T(u) - y_0(z)\|_Y \leq r(H_z) \leq \sup_{z \in \{S(B_\gamma, T(B_\gamma)) + \epsilon U\}} r(H_z) ,$$

so that

$$\bar{e} \leq \sup_{\substack{u \in B_\gamma \\ \|S(u, T(u)) - z\|_Z \leq \epsilon}} \|T(u) - \theta^*(z)\|_Y$$

$$\leq \sup_{z \in \{S(B_\gamma, T(B_\gamma)) + \epsilon U\}} r(H_z) .$$

Hence, (6.9) follows. $\qquad \square$

We remark that it is also possible to establish the existence of a Chebyshev center for the subset H_z under other (different) conditions, which we will not discuss any further.

§ 4. The Generalized Fock Space

The generalized (arbitrarily weighted) Fock space (GFS), denoted F, is a reproducing kernel Hilbert space of Volterra series referred to, in greater generality, Chapter 1. In this section, we focus attention on $F(E^N)$, the GFS of Volterra series defined on the N-dimensional Euclidean space E^N, which will be used in the description of general Nth-order nonlinear difference equations describing the input-output behavior of the nonlinear systems to be identified.

In order to provide insight into the system identification application, we define $F(E^N)$ by showing below how the members $f \in F(E^N)$ can be constructed from the components of E^N in terms of the following two properties, (a) and (b):

(a) f is a real-valued analytic function on a bounded set $\Omega \subset E^N$ defined by

$$\Omega = \{x \in E^N : ||x|| \leq \mu\}, \tag{6.10}$$

where μ is a positive constant. This implies that f can be represented by an N-variable power series, namely, a Volterra series in these N-variables, absolutely convergent at every $x \in \Omega$, expressible via

$$f(x) = \sum_{n=0}^{\infty} \frac{1}{n!} f_n(x), \tag{6.11}$$

where f_n is an n-power operator in x, given by

$$f_n(x) = \sum_{\substack{\sum_{k_1=0}^{n} \cdots \sum_{k_N=0}^{n} \\ |k|=k_1+k_2+\cdots k_N=n}} c_{k_1 \cdots k_N} \left(\frac{n!}{k_1! \cdots k_N!} \right) x_1^{k_1} \cdots x_N^{k_N}$$

$$:= \sum_{|k|=n} c_k \frac{n!}{k!} x^k,$$

where, in the last equality, we have used the notation

$$k = (k_1, \cdots, k_N)$$

$$|k| = \sum_{i=1}^{N} k_i$$

$$k! = k_1! \cdots k_N!$$

$$c_k = c_{k_1 \cdots k_N}$$

$$x^k = x_1^{k_1} \cdots x_N^{k_N},$$

and the constants c_k are symmetric with respect to their indices.

(b) For a sequence of given positive numbers

$$\rho = \{ \rho_0, \ \rho_1, \ \cdots \} \tag{6.12}$$

such that

$$\sum_{n=0}^{\infty} \frac{1}{\rho_n} \frac{\mu^{2n}}{n!} < \infty, \tag{6.13}$$

where some elements of ρ can be zero (if f belongs to a closed sub-space of $F(E^N)$), the f under consideration satisfies the following restrictions with respect to ρ:

$$\sum_{n=0}^{\infty} \frac{\rho_n}{n!} \sum_{|k|=n} \frac{n!}{k!} |c_k|^2 < \infty. \tag{6.14}$$

We are now ready to state the following result.

6.6. Theorem. *Under conditions* (6.10), (6.12), (6.13) *and* (6.14), *the completion of the linear space consisting of all the nonlinear functionals f defined by* (6.11), *endowed with the inner product and its induced norm defined below, is a reproducing kernel Hilbert space (RKHS), denoted also by $F(E^N)$:*

(i) *the inner product between any f and g in $F(E^N)$ is defined by*

$$\langle f, g \rangle_F = \sum_{n=0}^{\infty} \frac{\rho_n}{n!} \sum_{|k|=n} \frac{n!}{k!} c_k \, d_k, \tag{6.15}$$

where d_k are defined for g in the same way as c_k are defined for f, with the induced norm

$$\|f\| = \sqrt{\langle f, f \rangle_F}; \tag{6.16}$$

and

(ii) *the reproducing kernel $K(x, z)$ in $F(E^N)$ is given by*

$$K(x,y) = \tilde{K}(x^\top y) := \sum_{n=0}^{\infty} \frac{1}{n!}\frac{1}{\rho_n}(x^\top y)^n, \qquad \text{for all } x, y \in \Omega,$$

$$(6.17)$$

where

$$(x^\top y)^n := \sum_{|k|=n} \frac{n!}{k!} x^k y^k.$$

Here, we remark that the notation $\tilde{K}(x^\top y)$ will often be used, instead of $K(x, z)$, to emphasize the fact that the reproducing kernel of $F(E^N)$ is a function of the scalar variable $x^\top y$.

Proof. It follows from (6.14) that the series in (6.15) converges absolutely and, hence, (6.15) is well defined. It is easily verified that (6.15) indeed defines an inner product. The completeness of $F(E^N)$ in the norm (6.16), induced by this inner product, is established by imitating the proof of the completeness of l_2. To prove part (ii), note that, according to (6.13),

$$K(x, \cdot) \in F(E^N), \qquad \text{for all } x \in \Omega.$$

$$(6.18)$$

Furthermore, since

$$(x^\top y)^n = \sum_{|k|=n} \frac{n!}{k!} x^k y^k,$$

we have, according to (6.15) and (6.17),

$$\langle f(\cdot), K(x, \cdot) \rangle_F = \sum_{n=0}^{\infty} \frac{\rho_n}{n!} \sum_{|k|=n} \frac{n!}{k!} c_k \left(\frac{1}{\rho_n} x^k \right) = f(x), \qquad (6.19)$$

which shows that $K(x, \cdot)$, with $x \in \Omega$, is the representer of the point evaluation functional in $F(E^N)$. Positive definiteness of K follows from the positivity of the weights ρ_n. This, together with (6.18) and (6.19), establishes part (ii) of the theorem. $\square$

We remark that the following three special cases, depending on the nature of the weights ρ_n given in (6.12), are of interest:

(i) In the special case that the weights ρ_n are all equal to a constant, say ρ_0, the reproducing kernel K reduces to the exponential function

$$K(x,y) = \frac{1}{\rho_0} \exp(x^\top y).$$

(ii) If all the weights are zero except ρ_1, the elements of the closed subspace of $F(E^N)$, to which f belongs, are linear in x. $F(E^N)$ is then the N-dimensional ℓ^2-space, having the reproducing kernel

$$K(x,y) = \frac{1}{\rho_1} x^\top y.$$

(iii) If the weights up to ρ_m are the only nonzero components of ρ, then the elements f of the corresponding closed subspace of $F(E^N)$ are represented by finite power series up to degree m. This subspace is usually called the space of polynomic functionals.

In the following section we will assume all the weights ρ_n to be nonzero, where the results are applicable with minor modifications to the case in which the functionals f belong to a closed subspace of $F(E^N)$.

§ 5. Nonlinear System Identification Algorithms

As stated previously, an Nth-order nonlinear difference equation provides a generic model for relating the input-output samples of a nonlinear plant. Such input-output models in a stochastic context are known as ARMA (auto-regressive/moving-average) models and have, especially for the linear case, been widely discussed in the literature. The GFS framework developed in the previous section elicits a unique closed-form solution to the nonlinear case using a linear least squares approach.

It may be remarked that it is possible to identify nonlinear dynamical systems via state space models, but the derivation of state space models falls in the realm of realizability theory and, in general, state space realizations ought to be considered only after the input-output models such as the one described here are acquired from the data. For this reason we do not discuss here state space model identification and the underlying issues of controllability and observability associated with such models.

To begin with, we consider the single-input/single-output (SISO) case and show later how it can be extended to the multiple-input/multiple-output (MIMO) setting. We limit our attention to the AR model, relegating the case of the more general ARMA model to an exercise. Let the input-output description for an SISO system be represented by

$$y(k) + f(y(k-1), y(k-2), \cdots, y(k-N)) = u(k), \quad k \geq 0, \quad (6.20)$$

with given initial conditions

$$y(0) = y_0, \; y(-1) = y_1, \; \cdots, \; y(-N+1) = y_{N-1}.$$

Letting

$$y^{k-1} = \left(\, y(k-1), \; \cdots, \; y(k-N) \, \right)^{\top},$$

(6.20) can be abbreviated as

$$y(k) + f(y^{k-1}) = u(k). \tag{6.21}$$

We will require the following result on minimum norm interpolation in $F(E^N)$.

6.7. Theorem. *Let $v^1, \cdots, v^M \in E^N$ (called "domain training samples") be given vectors that are componentwise nonzero and distinct, namely, $v^i_j \neq v^i_m$ if $j \neq m$ for all i. Let also $z_1, \cdots, z_M$ (called the corresponding "range training samples") be given real numbers. Then there is a unique element $\hat{f}$ of minimum norm among all the $f \in F(E^N)$ that satisfy the interpolating constraints:*

$$f(v^i) = z_i, \quad i = 1, \cdots, M. \tag{6.22}$$

Moreover, $\hat{f}$ has the following expression:

$$\hat{f}(y) = \sum_{i=1}^{M} a_i K(v^i, y) = \sum_{i=1}^{M} a_i \tilde{K}((v^i)^{\top} y) \quad \text{for all } y \in E^N, \tag{6.23}$$

where $\tilde{K}((v^i)^{\top} \cdot)$, defined as in (6.17), are linearly independent elements of $F(E^N)$, and the parameter vector

$$a = (a_1, \cdots, a_M)^{\top}$$

is determined by

$$a = G^{-1}z, \tag{6.24}$$

in which

$$z = (z_1, \cdots, z_M)^{\top}$$

and G is the $M \times M$ matrix with elements

$$G_{ij} = \tilde{K}((v^j)^{\top} v^i). \tag{6.25}$$

Proof. Denote by S the closed subspace of $F(E^N)$ spanned by $K(v^i, \cdot)$, $i = 1, \cdots, M$, and let $S^{\perp}$ be the orthogonal complement of S. All elements f of $F(E^N)$ satisfying the interpolating constraints (6.22) lie in the hyperplane

$$H = f_0 + S^{\perp}, \tag{6.26}$$

where f_0 is an arbitrary element of S. By translating space H by $-f_0$, we conclude that the point (vector) $\hat{f}$ on H of minimum distance from the origin (i.e., of minimum norm) is mapped into the point f^* on $S^{\perp}$ of minimum distance from $-f_0$. Since $S^{\perp}$ is a closed subspace of $F(E^N)$, it follows from the last statement that f^* is the orthogonal projection of $-f_0$ on $S^\perp$ and hence

$$f^* - (-f_0) \in (S^\perp)^\perp = S, \qquad (6.27)$$

the latter equality following from the fact that S is closed. Also, by construction,

$$f^* - (-f_0) = f^* + f_0 = \hat{f}. \qquad (6.28)$$

It thus follows from (6.27) and (6.28) that $\hat{f} \in S$. Hence, it can be expressed in the form (6.23).

To prove that $\tilde{K}(v^{i\,T}\cdot)$, $i = 1, \cdots, M$, are linearly independent, assume otherwise. Then there are constants c_i, $i = 1, \cdots, M$, some of which are nonzero, such that

$$\sum_{i=1}^{M} c_i \tilde{K}(v^{i\,T} y) = 0, \qquad \text{for all } y \in E^N . \qquad (6.29)$$

Partially differentiating (6.29) M times, in succession, with respect to y, and then setting y equal to zero, we obtain

$$\frac{1}{\rho_1} c_1 v_j^1 + \frac{1}{\rho_1} c_2 v_j^2 + \cdots + \frac{1}{\rho_1} c_M v_j^M = 0,$$

$$\frac{1}{\rho_2} c_1 (v_j^1)^2 + \frac{1}{\rho_2} c_2 (v_j^2)^2 + \cdots + \frac{1}{\rho_2} c_M (v_j^M)^2 = 0,$$

$$\vdots$$

$$\frac{1}{\rho_M} c_1 (v_j^1)^M + \frac{1}{\rho_M} c_2 (v_j^2)^M + \cdots + \frac{1}{\rho_M} c_M (v_j^M)^M = 0.$$

These equations have a nonzero solution for the constants c_i, $i = 1, \cdots, M$, if and only if the determinant of the coefficients of c_i, $i = 1, \cdots, M$, vanishes. This determinant is that of the matrix

$$\frac{1}{\rho_0 \rho_1 \cdots \rho_M} \begin{bmatrix} v_j^1 & v_j^2 & \cdots & v_j^M \\ \vdots & \vdots & & \vdots \\ (v_j^1)^M & (v_j^2)^M & \cdots & (v_j^M)^M \end{bmatrix} . \qquad (6.30)$$

However, the constants $\rho_1, \cdots, \rho_M$ are nonzero, and, since the v_j^i, $i = 1, \cdots, M$, are nonzero and all distinct, the determinant of the matrix in (6.30), which is a Vandermonde matrix, cannot vanish. Hence, the constants c_i, $i = 1, \cdots, M$, in (6.29) must all vanish, and thus $\tilde{K}((v^i)^{\mathsf{T}} \cdot)$, $i = 1, \cdots, M$, are linearly independent elements of $F(E^N)$. Consequently, the Grammian G of these elements is nonsingular, and hence (6.24) follows. Finally, we point out that the expression (6.25) for the elements G_{ij} of this matrix follows directly from (6.19) by replacing $f(\cdot)$ by $K(v^i, \cdot)$ and $K(x, \cdot)$ by $K(v^j, \cdot)$ in that expression. $\qquad\square$

We next prove a concrete manifestation of Theorem 6.5. For this purpose, we first need the following concept.

6.8. Definition. A set Γ in a normed linear space is said to be *symmetric* if there exists an $f_c \in \Gamma$, called its *center*, with the property that

$$f_c + \eta \in \Gamma \quad \Longrightarrow \quad f_c - \eta \in \Gamma.$$

We can now establish the following result.

6.9. Theorem. *If Γ is a symmetric, closed and bounded subset of $F(E^N)$, then the center f_c of Γ minimizes the error*

$$e(f) = \sup_{\tilde{f}} \|f - \tilde{f}\| \tag{6.31}$$

over all $\tilde{f} \in \Gamma$.

Proof. Introduce, for convenience, the notation

$$g(f - \tilde{f}) := \|f - \tilde{f}\|.$$

Since g is a function of $f - \tilde{f}$ and Γ is symmetric, we may consider the translated set $\Gamma' = \Gamma - f_c$, which is symmetric about the origin. We thus reduce the proof of the theorem to the minimization of e on Γ'. Let $\{f_i\}_{i=0}^{\infty}$ be a sequence in Γ' such that $g(f_i) \to e(0)$ as $i \to \infty$. Then for every $\epsilon > 0$ there is an $n(\epsilon)$ such that

$$e(0) \leq g(f_i) + \epsilon, \quad i \geq n(\epsilon),$$

and, by the norm properties of g, for any $h \in \Gamma'$,

$$
\begin{aligned}
2g(f_i) = g(2f_i) &= g(f_i + h + f_i - h) \\
&\leq g(f_i + h) + g(f_i - h) \\
&= g(h - (-f_i)) + g(f_i - h) \,.
\end{aligned}
$$

We conclude that either (a) $g(h - f_i) \geq g(f_i)$ or (b) $g(h - (-f_i)) \geq g(f_i)$ is true. If (a) holds, then

$$
g(h - f_i) \geq g(f_i) \geq e(0) - \epsilon \,. \tag{6.32}
$$

On the other hand, if (b) is true, then

$$
g(h - (-f_i)) \geq g(f_i) \geq e(0) - \epsilon. \tag{6.33}
$$

Now (6.32) implies that

$$
e(0) \leq g(h - f_i) + \epsilon \leq \sup_{f \in \Gamma'} g(h - f) + \epsilon \tag{6.34}
$$

or, by (6.33),

$$
e(0) \leq g(h - (-f_i)) + \epsilon \leq \sup_{f \in \Gamma'} g(h - f) + \epsilon \,, \tag{6.35}
$$

since Γ' is symmetric about the origin, so that f_i and $-f_i$ are elements of Γ' for every i. Furthermore, since ϵ is arbitrary, (6.34) and (6.35) imply that $e(0) \leq e(h)$ for every $h \in \Gamma'$. This establishes the result of the theorem. □

Consider now the set

$$
\Lambda = B_\gamma \cap H \,, \tag{6.36}
$$

where H is the hyperplane defined by (6.26), and B_γ is a closed ball in $F(E^N)$ centered at the origin and with the radius γ sufficiently large so as Λ is nonempty. We have the following theorem.

6.10. Theorem. *The solution* (6.23) *to the minimum norm inter-polation problem minimizes the error* (6.31) *over* $\Gamma = \Lambda$.

We remark that in this min-max error criterion sense, $\hat{f}$ con-stitutes a "best approximation" to f given the data in the form indicated.

Proof. The set Λ is clearly closed, symmetric, and bounded, as required of the set Γ in Theorem 6.9.

We need to show that $\hat{f}$ defined by (6.23) is the center f_c of Γ, the result of the theorem then follows from Theorem 6.9. For this purpose, suppose that for any arbitrary η, $\hat{f} - \eta$ belongs to Γ and hence to H. Then, since $\hat{f} \in S$, η must belong to $S^\perp$, where, as before, S denotes the span of $K(v^i, \cdot)$, $i = 1, \cdots, M$. Hence $\hat{f} + \eta \in H$ and

$$
\begin{aligned}
||\hat{f} + \eta||^2 &= \langle \hat{f} + \eta, \hat{f} + \eta \rangle_F \\
&= ||\hat{f}||^2 + 2\langle \hat{f}, \eta \rangle_F + ||\eta||^2 \\
&= ||\hat{f}||^2 + ||\eta||^2 \\
&= ||\hat{f}||^2 - 2\langle \hat{f}, \eta \rangle_F + ||\eta||^2 \\
&= ||\hat{f} - \eta||^2 \le \gamma^2 ,
\end{aligned} \tag{6.37}
$$

where the second and third equalities follow from the orthogonality of $\hat{f}$ and η. Consequently, $\hat{f} + \eta \in \Lambda$ and so $\hat{f} = f_c$. $\square$

From elementary considerations, one can show that a bound on the approximation error is provided by

$$
e(\hat{f}) \le \gamma^2 - z^\top G^{-1} z , \tag{6.38}
$$

where the notation is as referred to in Theorems 6.7 and 6.10.

More generally, when the available number of the domain-range training sample pairs (v^j, z_j), $j = 1, \cdots, P$, is large com-pared to the desired dimension M of the approximation subspace in $F(E^N)$, one can obtain the best approximation $\hat{f}$ of f from the training data by smoothing rather than interpolation. To this end, after selecting the dimension M of the approximation subspace, one may choose any one of the standard clustering algorithms, available in the literature, to partition the domain training set $\{v^1, \cdots, v^P\}$ into M clusters $C_1, \cdots, C_M$. Let $\bar{v}^1, \cdots, \bar{v}^P \in E^N$ denote the means of $C_1, \cdots, C_M$, respectively, i.e.,

$$\bar{v}^i = \frac{1}{N_i} \sum_{v^k \in C_i} v^k, \qquad i = 1, \cdots, M \qquad (6.39)$$

where $N_i =$ number of elements in C_i. Then, the best approximation to f in $F(E^N)$ may be viewed as the $\hat{f}$ in the span of $\{K(\bar{v}^i, \cdot) : i = 1, \cdots, M\}$, the evaluations of which at v^j, $j = 1, \cdots, P$, are the closest to the corresponding outputs z^j, $j = 1, \cdots, P$, in the least squares sense. This leads to the following result.

6.11. Theorem. *Let $v^1, \cdots, v^P \in E^N$ be a set of domain training samples and the real numbers $z_1, \cdots, z_P$ the corresponding range training samples, from which the M-dimensional best approximation $\hat{f}$ needs to be recovered. Partition the set $\{v^1, \cdots, v^P\}$ into M clusters $C_1, \cdots, C_M$ according to some appropriate (unweighted or weighted) metric in E^N. Let the means $\bar{v}^1, \cdots, \bar{v}^M$ of these clusters be componentwise nonzero and distinct. Then, there is a unique $\hat{f}$ in $F(E^N)$ expressed by*

$$\hat{f}(y) = \sum_{i=1}^{M} a_i K(\bar{v}^i, y) = \sum_{i=1}^{M} a_i \tilde{K}((\bar{v}^i)^\top y), \qquad \text{for all } y \in E^N,$$

$$(6.40)$$

which minimizes

$$\min_{a_1, \cdots, a_M} \left| \sum_{j=1}^{P} \sum_{i=1}^{M} a_i K(\bar{v}^i, v^j) - z_j \right|^2. \qquad (6.41)$$

The parameter vector

$$a = (a_1, \cdots, a_M)^\top \qquad (6.42)$$

is determined from

$$a = (G^\top G)^{-1} G^\top z \qquad (6.43)$$

where

$$z = (z_1, \cdots, z_P)^\top \tag{6.44}$$

and G is a $P \times M$ matrix with elements

$$G_{ij} = \tilde{K}((\bar{v}^j)^\top v^i), \quad i = 1, \cdots, P, \quad j = 1, \cdots, M. \tag{6.45}$$

The proof is elementary and is therefore left as an exercise.

We remark that while Theorem 6.10 constitutes the best approximation of f in the min-max sense, Theorem 6.11 may be viewed as a best approximation in the least square sense. Both these theorems permit us to formulate the problem of identification of an Nth-order nonlinear SISO system described by (6.20) in the above two senses, from the input-output samples.

Specifically, consider a set of M input samples $\tilde{u}(k)$, $\tilde{u}(k+1)$, $\cdots$, $\tilde{u}(k+M-1)$ and a related set of $M + N$ output samples $\tilde{y}(k-N)$, $\tilde{y}(k-N+1)$, $\cdots$, $\tilde{y}(k)$, $\cdots$, $\tilde{y}(k+M-1)$. Implicit in our formulation is that the order N of the system is known *a priori*. If that is not the case (as revealed by a bound on approximation error such as (6.36)) a better fit to the system to be identified may be attempted at a second pass by increasing N and/or M. The system identification can be achieved by means of Algorithms 1 to 3 described below.

6.12. Algorithm 1 (Nonrecursive Interpolation). First, we arrange the input and output samples as follows:

$$\tilde{y}^{k+i-2} = (\tilde{y}(k+i-2), \tilde{y}(k+i-3), \cdots, \tilde{y}(k+i-N-1))^\top,$$
$$i = 1, 2, \cdots, M, \tag{6.46}$$

$$\tilde{z}(k+i-1) = \tilde{u}(k+i-1) - \tilde{y}(k+i-1), \quad i = 1, 2, \cdots, M. \tag{6.47}$$

Then, with the notational agreements

$$\tilde{y}^{k+i-2} = v^i, \quad i = 1, 2, \cdots, M, \tag{6.48}$$
$$\tilde{z}(k+i-1) = z_i, \quad i = 1, 2, \cdots, M, \tag{6.49}$$
$$y^{k-1} = y, \tag{6.50}$$

and under the condition that the outputs

$$\tilde{y}(k-N),\ \tilde{y}(k-N+1),\ \cdots,\ \tilde{y}(k),\ \cdots,\ \tilde{y}(k+M-1) \quad (6.51)$$

of the system permit the conditions on the components of v^i in Theorems 6.7 and 6.10 be satisfied, the expression (6.30) becomes the best approximation $\hat{f}$ to f in (6.20) in the sense of Theorem 6.10. Thus the objective of the algorithm is accomplished.

6.13. Algorithm 2 (Recursive Interpolation) If the input/output samples are provided sequentially to the designer, then Algorithm 1 can be converted to a recursive form by propagating the inverse G^{-1} in (6.24) recursively. Thus let $G(M)$ denote G when it is an $M \times M$ matrix. Then by a standard derivation (the "matrix inversion lemma") one can relate $G^{-1}(M+1)$ to $G^{-1}(M)$ by

$$G^{-1}(M+1) = \begin{bmatrix} G^{-1}(M) & c^{M+1} \\ (c^{M+1})^\top & \alpha_{M+1}^{-1} \end{bmatrix}, \quad (6.52)$$

where

$$c^{M+1} = G^{-1}(M)b^{M+1}, \quad (6.53)$$

$$b^{M+1} = (\tilde{K}((v^1)^\top v^{M+1}), \cdots, \tilde{K}((v^M)^\top v^{M+1}))^\top, \quad (6.54)$$

$$\alpha = \tilde{K}((v^{(M+1)})^\top v^{M+1}), \quad (6.55)$$

$$c^{M+1} = G^{-1}(M)b^{M+1}. \quad (6.56)$$

6.14. Algorithm 3 (Nonrecursive Smoothing) Suppose now that we wish to fit a large number P of samples to a model belonging to a subspace of $F(E^N)$ of dimension $M < P$. In this case, we replace M by P in equations (6.46) to (6.49). Then, with the notational agreements (6.48) to (6.50) (with M replaced by P) we apply the results of Theorem 6.11, under the conditions stated in that theorem, in order to obtain a least squares approximation $\hat{f}$ to the functional f in (6.20).

We remark that the pseudo-inverse in (6.43) can be propagated recursively as in Algorithm 2 using the matrix inversion lemma. We leave the details as an exercise.

Finally, the results presented above can be extended to the multi-input/multi-output (MIMO) case. The approach is summarized in Algorithm 4.

6.15. Algorithm 4 (MIMO) Denote by $u_1(k), \cdots, u_p(k)$ the inputs and by $y_1(k), \cdots, y_q(k)$ the outputs of the system. Then for each output $y_i(k)$ we may write a generic nonlinear difference equation of the form

$$y_i(k) + f_i(y_1^{k-1}, \cdots, y_q^{k-1}, u_1(k), \cdots, u_p(k)) = 0, \quad i = 1, \cdots, q,$$
(6.57)

where, in a way analogous to (6.21), we have used the notation

$$y_i^{k-1} = (y_i(k-1), \cdots, y_i(k - N_i))^\top, \quad i = 1, \cdots, q,$$
(6.58)

with N_i denoting the order of the difference equation pertaining to the output y_i.

We can now assume that the functionals f_i, $i = 1, \cdots, q$, belong to $F(E^{N_1 + \cdots + N_q + p})$. Suppose that we are given a set of inputs $\tilde{u}_m(k+i)$, $m = 1, \cdots, p$, $i \in S_1$, and a set of corresponding outputs $\tilde{y}_n(k+j)$, $n = 1, \cdots, q$, $j \in S_2$, where S_1 and S_2 are appropriate sets of integers. Then the best approximations $\hat{f}_i$ for the functionals f_i can be obtained in the same way as for the SISO case, and thus the nonlinear MIMO system identification algorithm, in nonrecursive or recursive form, can be implemented in a way analogous to that for the SISO case. We omit the details for the simplicity of notation and presentation.

6.16. Example. Consider the nonlinear system described by

$$y(k) - \sinh\{y(k-1)\} = u(k).$$
(6.59)

If we choose the GFS $F(E^1)$ with all the weights ρ_n set equal to 1, then from very few input-output samples, an exact approximation for the "system" (with zero input)

$$y(k) = f(y^{k-1}) \quad \text{or} \quad f(\cdot(k-1)) = \sinh\{\cdot(k-1)\}$$
(6.60)

can be obtained as

$$\hat{f}(\cdot(k-1)) = \frac{1}{2}e^{(\cdot)(k-1)} - \frac{1}{2}e^{-(\cdot)(k-1)},$$
(6.61)

or more precisely,

$$\hat{f}(y(k-1)) = \frac{1}{2}e^{y(k-1)} - \frac{1}{2}e^{-y(k-1)}.$$
(6.62)

§ 6. Extension to Continuous-Time-Parameter Systems

The approach to nonlinear system identification described above can be extended to those nonlinear systems that have input-output relations described by an Nth-order nonlinear differential equation of the form

$$y^{(N)}(t) + f(y^{(N-1)}(t), \cdots, y^{(1)}(t), y(t)) = u(t), \quad t \geq 0, \quad (6.63)$$

where $f \in F(E^N)$ and $(\cdot)^{(k)} = \frac{d^k}{dt^k}(\cdot)$, under prescribed initial conditions

$$y(0) = y_0, \; y^{(1)}(0) = y_1, \; \cdots, \; y^{(N-1)}(0) = y_{N-1}. \quad (6.64)$$

From the values of y and its derivatives up to order $N - 1$ and the corresponding inputs at time instants $t_1 < t_2 < \cdots < t_M$, we may obtain equations of the form (6.22) under the notational agreements

$$v^i = (y^{(N-1)}(t_i), \cdots, y^{(1)}(t_i), y(t_i))^\top \quad (6.65)$$
$$z_i = u(t_i) - y^{(N)}(t_i). \quad (6.66)$$

Clearly, with this understanding all the preceding algorithms can be applied for obtaining a best approximation for f. However, a difficulty arises in pursuing this approach when the data is noisy. In this case, conventional derivatives have to be replaced by distributional derivatives. Another avenue that one may wish to pursue is to filter noise out before the current approach is applied.

In summary, we have shown how RKHS techniques, which have been successfully used in solving linear problems, provide a powerful framework for the solution of the nonlinear system identification problem. Recently, these concepts have made their way into the modeling and analysis of artificial neural networks as embodied in some of the references listed at the end of this chapter.

Exercises

6.1. Give nontrivial examples to show that the lower bound for the identification error given in (6.3) is sharp, in the sense that it can be achieved by some nonlinear operators, and hence cannot be improved.

6.2. Is the upper bound for the identification error given in (6.4) also sharp? Prove or disprove your claim.

6.3. Give examples of a Chebyshev center and radius of a set in a Banach space. In particular, give examples to show that a Chebyshev center of a set is not necessarily an element of the set and may not be unique in general. Also, give examples to show that if the space to which the set belongs is not Banach, then a Chebyshev center of the set may not exist, even if the set is closed, bounded, and convex.

6.4. Under different conditions, establish some other existence results similar to Theorem 6.5.

6.5. Supply a detailed verification for Theorem 6.14.

6.6. Give two examples of continuous linear functionals on $F(E^N)$ by exhibiting expressions for these functionals.

6.7. What are the representers for the above functionals in $F(E^N)$?

6.8. Explain how Theorem 6.7 can be extended to the case in which the constraints are established by M arbitrary continuous linear functionals on $F(E^N)$ rather than by the point evaluation functionals (6.22).

6.9. Extend the results presented in Section 5 to the nonlinear ARMA model.

6.10. Explain how the results of Section 5 can be applied to the identification of a system in which a continuous input is sent to the value at time t of the output by a Volterra series of the type described in Section 9 of Chapter 1. Hint: see [19].

References

[1] N. Aronszajn, "Theory of reproducing kernels", *Trans. Amer. Math. Soc.*, **68** (1950), 337–404.

[2] G. Chen, "On the information-based complexity of optimal reconstruction for nonlinear systems," *Sys. Contr. Lett.*, **14** (1990), 161–167.

[3] G. Chen, "Optimal recovery of certain nonlinear analytic mappings," in *Optimal Recovery* (B. Bojanov and H. Woźniakowski, eds.), pp. 141–144, Nova Sci. Pub. Inc., Commack, NY, 1992.

[4] R.J.P.de Figueiredo, "A generalized Fock space framework for nonlinear system and signal analysis," *IEEE Trans. Circ. Sys.*, **30** (1983), 637–647.

[5] R.J.P.de Figueiredo, "A new nonlinear analytic framework for modeling artificial neural networks" (invited paper), *Proc. of IEEE Int'l Sympos. Circ. Sys.*, New Orleam, LA, May 1-3 (1990), pp. 723–726.

[6] R.J.P.de Figueiredo, "Mathematical foundations of optimal interpolative neural networks", in *Artificial Intelligence, Expert Systems, and Symbolic Computing*, (E.N. Houstis and J.R. Rice, eds.) pp. 303–319, Elsevier Sci. Pub., Amsterdam, The Netherlands, 1992.

[7] R.J.P.de Figueiredo, A. Caprihan, and A.N. Netravali, "On optimal modeling of systems," *J. Optmiz. Theory*, **11** (1973), 68–83.

[8] R.J.P.de Figueiredo and G. Chen, "Optimal interpolation in a generalized Fock space of analytic functions," in *Approximation Theory VI* (C.K. Chui *et al.*, eds.), pp. 247–250, Academic Press, New York, 1989.

[9] R.J.P.de Figueiredo and T.A.W. Dwyer, "A best approximation framework and implementation for simulation of large-scale nonlinear systems," *IEEE Trans. Circ. Sys.*, **27** (1980), 1005–1014.

[10] A.L. Garkavi, "The best possible net and the best possible cross-section of a set in a normed space," *Amer. Math. Soc. Transl.*, **39** (1964), 111–132.

[11] G.C. Goodwin and K.S. Sin, *Adaptive Filtering Prediction and Control*, Prentice-Hall, Englewood Cliffs, N. J., 1984.

[12] L.T. Grujić and A.N. Michel, "Exponential stability and trajectory bounds of neural networks under structural variations," *IEEE Trans. Circ. Sys.*, **38** (1991), 1182-1192.

[13] L. Ljung, *System Identification: Theory for the User*, Prentice-Hall, Englewood Cliffs, N. J., 1987.

[14] C.A. Micchelli and T.J. Rivlin, "A survey of optimal recovery," in *Optimal Estimation in Approximation Theory* (C.A. Micchelli and T.J.Rivlin, eds.), pp. 1−54, Plenum Press, New York, 1977.

[15] A.N. Netravali and R.J.P.de Figueiredo, "On the identification of nonlinear dynamical system," *IEEE Trans. Auto. Contr.* **16** (1971), 28−35.

[16] M. Reed and B. Simon, *Methods of Modern Mathematical Physics*, Vol.1: *Functional Analysis*, Academic Press, New York, 1973.

[17] S.K. Sin and R.J.P.de Figueiredo, "Efficient learning procedures for optimal interpolative nets," *Neural Networks*, **6** (1993), 99−113.

[18] J.F. Traub, G.W. Wasilkowski, and H. Woźniakowski, *Information-Based Complexity*, Academic Press, New York, 1988.

[19] L.V. Zyla and R.J.P.de Figueiredo, "Nonlinear system identification based on a Fock space framework," *SIAM J. Contr. Optimiz.*, **21** (1983), 931−939.

Index

219